Principles of Cosmology

Principles of Cosmology

Callen Hogan

RCALLISTO
REFERENCE

www.callistoreference.com

Callisto Reference,
118-35 Queens Blvd., Suite 400,
Forest Hills, NY 11375, USA

Visit us on the World Wide Web at:
www.callistoreference.com

ISBN: 978-1-64116-230-2 (Hardback)

Cataloging-in-Publication Data

Principles of cosmology / Callen Hogan.
    p. cm.
Includes bibliographical references and index.
ISBN 978-1-64116-230-2
1. Cosmology. 2. Universe. I. Hogan, Callen.
QB981 .P75 2019
523.1--dc23

Table of Contents

Preface

The field of cosmology investigates the origin, evolution and the future of the universe as well as the scientific principles and laws that guide these phenomena. The fields of physics and astrophysics use mathematical formulations, scientific observation and experimentation for the development of this subject. The universe is believed to have started with the big bang, after which it underwent a process of cosmic inflation. Modern cosmology evolved after the formulation of Einstein's general theory of relativity. Observations of the cosmic microwave background radiation, new galaxy redshift surveys, gravitational lensing, observations of supernovae, etc. have further advanced the understanding of the universe. This book is a valuable compilation of topics that covers the most fundamental theories and principles of cosmology. It presents the complex subject of cosmology in the most comprehensible and easy to understand language. This book is meant for students who are looking for an elaborate reference text in this domain.

To facilitate a deeper understanding of the contents of this book a short introduction of every chapter is written below:

Chapter 1- The universe comprises of space and time, large celestial bodies, such as stars, planets and galaxies as well as all of matter and energy. This is an introductory chapter, which will introduce briefly all the significant aspects of the universe, such as chronology of the universe, observable universe, age of the universe and shape of the universe.

Chapter 2- The study of the origin, evolution and the eventual fate of the universe is under the domain of cosmology. This chapter discusses in detail the theories and models central to the development of cosmology, such as Big Bang theory, Lambda-CDM model, cosmological constant and recombination, among others.

Chapter 3- The study of the large-scale structures of the universe, and their dynamics is under the scope of physical cosmology. This chapter has been carefully written to provide an easy understanding of physical cosmology through the principal topics like cosmological principle, Hubble's law, baryogenesis and cosmological perturbation theory.

Chapter 4- The cosmic background radiation is the electromagnetic radiation from the event of the big bang. The origin of the radiation is dependent on the spectrum region under observation. This chapter discusses in detail the background radiation of the universe such as the cosmic microwave background, cosmic infrared background and cosmic neutrino background.

Chapter 5- Dark matter is a hypothetical form of matter that constitutes approximately 80% of the entire matter in the universe. The topics elaborated in this chapter include various significant topics such as cold dark matter, hot dark matter, mixed dark matter and dark matter halo, which will provide an in-depth understanding of dark matter in the universe.

I would like to share the credit of this book with my editorial team who worked tirelessly on this book. I owe the completion of this book to the never-ending support of my family, who supported me throughout the project.

Callen Hogan

Chapter 1

Understanding the Universe

The universe comprises of space and time, large celestial bodies, such as stars, planets and galaxies as well as all of matter and energy. This is an introductory chapter, which will introduce briefly all the significant aspects of the universe, such as chronology of the universe, observable universe, age of the universe and shape of the universe.

The Universe is everything we can touch, feel, sense, measure or detect. It includes living things, planets, stars, galaxies, dust clouds, light, and even time. Before the birth of the Universe, time, space and matter did not exist.

The Universe contains billions of galaxies, each containing millions or billions of stars. The space between the stars and galaxies is largely empty. However, even places far from stars and planets contain scattered particles of dust or a few hydrogen atoms per cubic centimeter. Space is also filled with radiation (e.g. light and heat), magnetic fields and high energy particles (e.g. cosmic rays).

The Universe is incredibly huge. It would take a modern jet fighter more than a million years to reach the nearest star to the Sun. Travelling at the speed of light (300,000 km per second), it would take 100,000 years to cross our Milky Way galaxy alone.

The universe was born with the Big Bang as an unimaginably hot, dense point. When the universe was just 10^{-34} of a second or so old — that is, a hundredth of a billionth of a trillionth of a trillionth of a second in age — it experienced an incredible burst of expansion known as inflation, in which space itself expanded faster than the speed of light. During this period, the universe doubled in size at least 90 times, going from subatomic-sized to golf-ball-sized almost instantaneously.

During the first three minutes of the universe, the light elements were born during a process known as Big Bang nucleosynthesis. Temperatures cooled from 100 nonillion (10^{32}) Kelvin to 1 billion (10^9) Kelvin, and protons and neutrons collided to make deuterium, an isotope of hydrogen. Most of the deuterium combined to make helium, and trace amounts of lithium were also generated.

The Big Bang

The Big Bang did not occur as an explosion in the usual way one think about such things, despite one might gather from its name. The universe did not expand into space, as space did not exist before the universe, according to NASA Instead, it is better to think of the Big Bang as the simultaneous appearance of space everywhere in the universe. The universe has not expanded from any one spot since the Big Bang — rather, space itself has been stretching, and carrying matter with it.

Since the universe by its definition encompasses all of space and time as we know it, NASA says it is beyond the model of the Big Bang to say what the universe is expanding into or what gave rise to the Big Bang. Although there are models that speculate about these questions, none of them have made realistically testable predictions as of yet.

In 2014, scientists from the Harvard-Smithsonian Center for Astrophysics announced that they had found a faint signal in the cosmic microwave background that could be the first direct evidence of gravitational waves, themselves considered a "smoking gun" for the Big Bang. The findings were hotly debated, and astronomers soon retracted their results when they realized dust in the Milky Way could explain their findings.

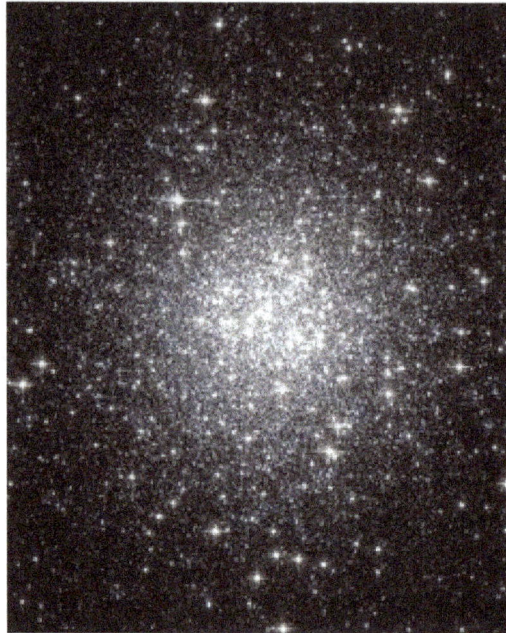

The globular cluster NGC 6397 contains around 400,000 stars and is located about 7,200 light years away in the southern constellation Ara. With an estimated age of 13.5 billion years, it is likely among the first objects of the Galaxy to form after the Big Bang.

Physical Properties

Of the four fundamental interactions, gravitation is the dominant at astronomical length scales. Gravity's effects are cumulative; by contrast, the effects of positive and negative charges tend to cancel one another, making electromagnetism relatively insignificant on astronomical length scales. The remaining two interactions, the weak and strong nuclear forces, decline very rapidly with distance; their effects are confined mainly to sub-atomic length scales.

The Universe appears to have much more matter than antimatter, an asymmetry possibly related to the CP violation. This imbalance between matter and antimatter is partially responsible for the existence of all matter existing today, since matter and antimatter, if equally produced at the Big Bang, would have completely annihilated each other and leaved only photons as a result of their interaction. The Universe also appears to have neither net momentum nor angular momentum, which follows accepted physical laws if the Universe is finite. These laws are the Gauss's law and the non-divergence of the stress-energy-momentum pseudotensor.

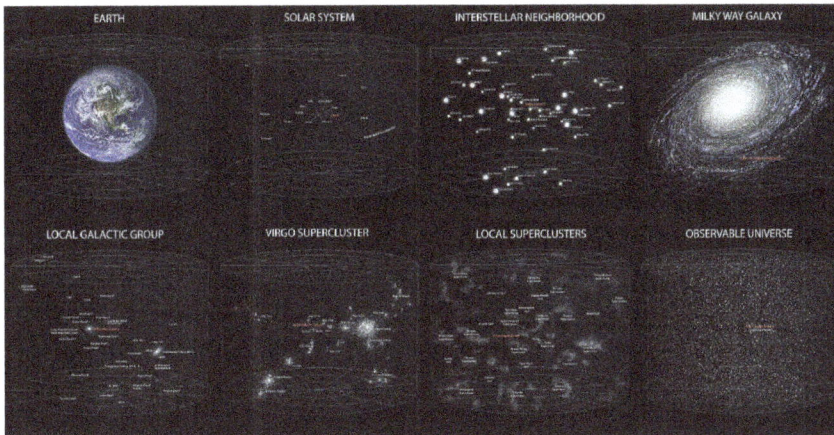

This diagram shows Earth's location in the Universe.

Size and Regions

The size of the Universe is somewhat difficult to define. According to the general theory of relativity, some regions of space may never interact with ours even in the lifetime of the Universe due to the finite speed of light and the ongoing expansion of space. For example, radio messages sent from Earth may never reach some regions of space, even if the Universe were to exist forever: space may expand faster than light can traverse it.

Distant regions of space are assumed to exist and to be part of reality as much as we are, even though we can never interact with them. The spatial region that we can affect and be affected by is the observable universe. The observable universe depends on the location of the observer. By traveling, an observer can come into contact with a greater region of spacetime than an observer who remains still. Nevertheless, even the most rapid traveler will not be able to interact with all of space. Typically, the observable universe is taken to mean the portion of the Universe that is observable from our vantage point in the Milky Way.

The proper distance—the distance as would be measured at a specific time, including the present—between Earth and the edge of the observable universe is 46 billion light-years (14 billion parsecs), making the diameter of the observable universe about 91 billion light-years (28×10^9 pc). The distance the light from the edge of the observable universe has travelled is very close to the age of the Universe times the speed of light, 13.8 billion light-years (4.2×10^9 pc), but this does not represent the distance at any given time because the edge of the observable universe and the Earth have since moved further apart. For comparison, the diameter of a typical galaxy is 30,000 light-years (9,198 parsecs), and the typical distance between two neighboring galaxies is 3 million light-years (919.8 kiloparsecs). As an example, the Milky Way is roughly 100,000–180,000 light years in diameter, and the nearest sister galaxy to the Milky Way, the Andromeda Galaxy, is located roughly 2.5 million light years away.

Because we cannot observe space beyond the edge of the observable universe, it is unknown whether the size of the Universe in its totality is finite or infinite. Estimates for the total size of the universe, if finite, reach as high as megaparsecs, implied by one resolution of the No-Boundary Proposal.

Age and Expansion

Astronomers calculate the age of the Universe by assuming that the Lambda-CDM model accurately describes the evolution of the Universe from a very uniform, hot, dense primordial state to its present state and measuring the cosmological parameters which constitute the model. This model is well understood theoretically and supported by recent high-precision astronomical observations such as WMAP and Planck. Commonly, the set of observations fitted includes the cosmic microwave background anisotropy, the brightness/redshift relation for Type Ia supernovae, and large-scale galaxy clustering including the baryon acoustic oscillation feature. Other observations, such as the Hubble constant, the abundance of galaxy clusters, weak gravitational lensing and globular cluster ages, are generally consistent with these, providing a check of the model, but are less accurately measured at present. Assuming that the Lambda-CDM model is correct, the measurements of the parameters using a variety of techniques by numerous experiments yield a best value of the age of the Universe as of 2015 of 13.799 ± 0.021 billion years.

Over time, the Universe and its contents have evolved; for example, the relative population of quasars and galaxies has changed and space itself has expanded. Due to this expansion, scientists on Earth can observe the light from a galaxy 30 billion light years away even though that light has traveled for only 13 billion years; the very space between them has expanded. This expansion is consistent with the observation that the light from distant galaxies has been redshifted; the photons emitted have been stretched to longer wavelengths and lower frequency during their journey. Analyses of Type Ia supernovae indicate that the spatial expansion is accelerating.

The more matter there is in the Universe, the stronger the mutual gravitational pull of the matter. If the Universe were *too* dense then it would re-collapse into a gravitational singularity. However, if the Universe contained too *little* matter then the expansion would accelerate too rapidly for planets and planetary systems to form. Since the Big Bang, the universe has expanded monotonically. Perhaps unsurprisingly, our universe has just the right mass density of about 5 protons per cubic meter which has allowed it to expand for the last 13.8 billion years, giving time to form the universe as observed today.

There are dynamical forces acting on the particles in the Universe which affect the expansion rate. Before 1998, it was expected that the rate of increase of the Hubble Constant would be decreasing as time went on due to the influence of gravitational interactions in the Universe, and thus there is an additional observable quantity in the Universe called the deceleration parameter which cosmologists expected to be directly related to the matter density of the Universe. In 1998, the deceleration parameter was measured by two different groups to be consistent with −1 but not zero, which implied that the present-day rate of increase of the Hubble Constant is increasing over time.

Spacetime

Spacetimes are the arenas in which all physical events take place. The basic elements of spacetimes are events. In any given spacetime, an event is defined as a unique position at a unique time. A spacetime is the union of all events (in the same way that a line is the union of all of its points), formally organized into a manifold.

The Universe appears to be a smooth spacetime continuum consisting of three spatial dimensions and one temporal (time) dimension (an event in the spacetime of the physical Universe can therefore be identified by a set of four coordinates: (x, y, z, t)). On the average, space is observed to be very nearly flat (with a curvature close to zero), meaning that Euclidean geometry is empirically true with high accuracy throughout most of the Universe. Spacetime also appears to have a simply connected topology, in analogy with a sphere, at least on the length-scale of the observable Universe. However, present observations cannot exclude the possibilities that the Universe has more dimensions (which is postulated by theories such as the String theory) and that its spacetime may have a multiply connected global topology, in analogy with the cylindrical or toroidal topologies of two-dimensional spaces. The spacetime of the Universe is usually interpreted from a Euclidean perspective, with space as consisting of three dimensions, and time as consisting of one dimension, the "fourth dimension". By combining space and time into a single manifold called Minkowski space, physicists have simplified a large number of physical theories, as well as described in a more uniform way the workings of the Universe at both the supergalactic and subatomic levels.

Spacetime events are not absolutely defined spatially and temporally but rather are known to be relative to the motion of an observer. Minkowski space approximates the Universe without gravity; the pseudo-Riemannian manifolds of general relativity describe spacetime with matter and gravity.

Shape

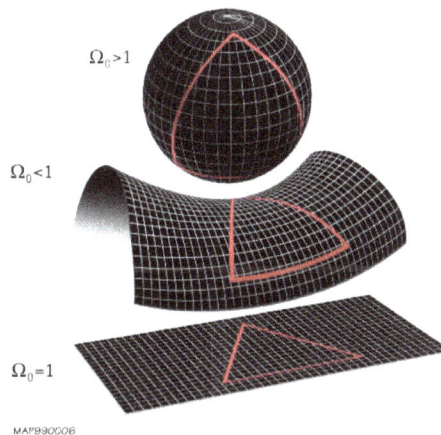

The three possible options for the shape of the Universe

General relativity describes how spacetime is curved and bent by mass and energy (gravity). The topology or geometry of the Universe includes both local geometry in the observable universe and global geometry. Cosmologists often work with a given space-like slice of spacetime called the comoving coordinates. The section of spacetime which can be observed is the backward light cone, which delimits the cosmological horizon. The cosmological horizon (also called the particle horizon or the light horizon) is the maximum distance from which particles can have traveled to the observer in the age of the Universe. This horizon represents the boundary between the observable and the unobservable regions of the Universe. The existence, properties, and significance of a cosmological horizon depend on the particular cosmological model.

An important parameter determining the future evolution of the Universe theory is the density parameter, Omega (Ω), defined as the average matter density of the universe divided by a critical value of that density. This selects one of three possible geometries depending on whether Ω is equal to, less than, or greater than 1. These are called, respectively, the flat, open and closed universes.

Observations, including the Cosmic Background Explorer (COBE), Wilkinson Microwave Anisotropy Probe (WMAP), and Planck maps of the CMB, suggest that the Universe is infinite in extent with a finite age, as described by the Friedmann–Lemaître–Robertson–Walker (FLRW) models. These FLRW models thus support inflationary models and the standard model of cosmology, describing a flat, homogeneous universe presently dominated by dark matter and dark energy.

Support of Life

The Universe may be *fine-tuned*; the Fine-tuned Universe hypothesis is the proposition that the conditions that allow the existence of observable life in the Universe can only occur when certain universal fundamental physical constants lie within a very narrow range of values, so that if any of several fundamental constants were only slightly different, the Universe would have been unlikely to be conducive to the establishment and development of matter, astronomical structures, elemental diversity, or life as it is understood. The proposition is discussed among philosophers, scientists, theologians, and proponents of creationism.

Composition

The Universe is composed almost completely of dark energy, dark matter, and ordinary matter. Other contents are electromagnetic radiation (estimated to constitute from 0.005% to close to 0.01% of the total mass of the Universe) and antimatter.

The proportions of all types of matter and energy have changed over the history of the Universe. The total amount of electromagnetic radiation generated within the universe has decreased by 1/2 in the past 2 billion years. Today, ordinary matter, which includes atoms, stars, galaxies, and life, accounts for only 4.9% of the contents of the Universe. The present overall density of this type of matter is very low, roughly 4.5×10^{-31} grams per cubic centimetre, corresponding to a density of the order of only one proton for every four cubic meters of volume. The nature of both dark energy and dark matter is unknown. Dark matter, a mysterious form of matter that has not yet been identified, accounts for 26.8% of the cosmic contents. Dark energy, which is the energy of empty space and is causing the expansion of the Universe to accelerate, accounts for the remaining 68.3% of the contents.

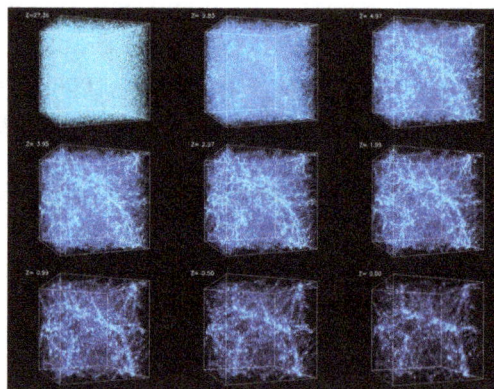

The formation of clusters and large-scale filaments in the cold dark matter model with dark energy. The frames show the evolution of structures in a 43 million parsecs (or 140 million light years) box from redshift of 30 to the present epoch (upper left z=30 to lower right z=0).

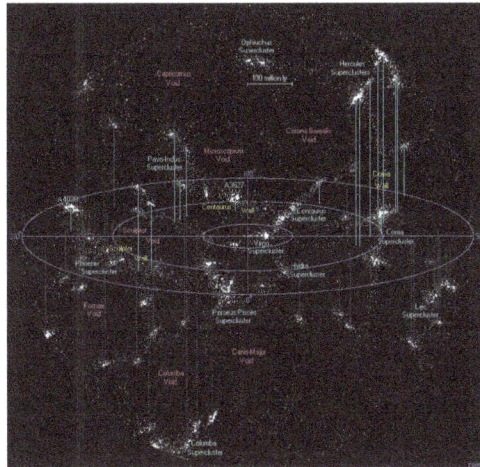

A map of the superclusters and voids nearest to Earth

Matter, dark matter, and dark energy are distributed homogeneously throughout the Universe over length scales longer than 300 million light-years or so. However, over shorter length-scales, matter tends to clump hierarchically; many atoms are condensed into stars, most stars into galaxies, most galaxies into clusters, superclusters and, finally, large-scale galactic filaments. The observable Universe contains approximately 300 sextillion (3×10^{23}) stars and more than 100 billion (10^{11}) galaxies. Typical galaxies range from dwarfs with as few as ten million (10^7) stars up to giants with one trillion (10^{12}) stars. Between the larger structures are voids, which are typically 10–150 Mpc (33 million–490 million ly) in diameter. The Milky Way is in the Local Group of galaxies, which in turn is in the Laniakea Supercluster. This supercluster spans over 500 million light years, while the Local Group spans over 10 million light years. The Universe also has vast regions of relative emptiness; the largest known void measures 1.8 billion ly (550 Mpc) across.

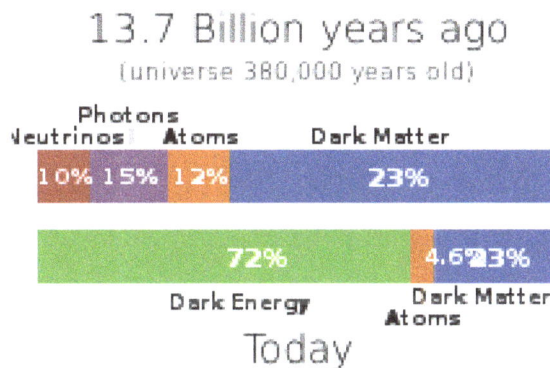

Comparison of the contents of the Universe today to 380,000 years after the Big Bang as measured with 5 year WMAP data. (Due to rounding errors, the sum of these numbers is not 100%).

The observable Universe is isotropic on scales significantly larger than superclusters, meaning that the statistical properties of the Universe are the same in all directions as observed from Earth. The Universe is bathed in highly isotropic microwave radiation that corresponds to a thermal equilibrium blackbody spectrum of roughly 2.72548 kelvin. The hypothesis that the large-scale Universe is homogeneous and isotropic is known as the cosmological principle. A Universe that is both homogeneous and isotropic looks the same from all vantage points and has no center.

Dark Energy

An explanation for why the expansion of the Universe is accelerating remains elusive. It is often attributed to "dark energy", an unknown form of energy that is hypothesized to permeate space. On a mass–energy equivalence basis, the density of dark energy ($\sim 7 \times 10^{-30}$ g/cm^3) is much less than the density of ordinary matter or dark matter within galaxies. However, in the present dark-energy era, it dominates the mass–energy of the universe because it is uniform across space.

Two proposed forms for dark energy are the cosmological constant, a *constant* energy density filling space homogeneously, and scalar fields such as quintessence or moduli, *dynamic* quantities whose energy density can vary in time and space. Contributions from scalar fields that are constant in space are usually also included in the cosmological constant. The cosmological constant can be formulated to be equivalent to vacuum energy. Scalar fields having only a slight amount of spatial inhomogeneity would be difficult to distinguish from a cosmological constant.

Dark Matter

Dark matter is a hypothetical kind of matter that is invisible to the entire electromagnetic spectrum, but which accounts for most of the matter in the Universe. The existence and properties of dark matter are inferred from its gravitational effects on visible matter, radiation, and the large-scale structure of the Universe. Other than neutrinos, a form of hot dark matter, dark matter has not been detected directly, making it one of the greatest mysteries in modern astrophysics. Dark matter neither emits nor absorbs light or any other electromagnetic radiation at any significant level. Dark matter is estimated to constitute 26.8% of the total mass–energy and 84.5% of the total matter in the Universe.

Ordinary Matter

The remaining 4.9% of the mass–energy of the Universe is ordinary matter, that is, atoms, ions, electrons and the objects they form. This matter includes stars, which produce nearly all of the light we see from galaxies, as well as interstellar gas in the interstellar and intergalactic media, planets, and all the objects from everyday life that we can bump into, touch or squeeze. As a matter of fact, the great majority of ordinary matter in the universe is unseen, since visible stars and gas inside galaxies and clusters account for less than 10 per cent of the ordinary matter contribution to the mass-energy density of the universe.

Ordinary matter commonly exists in four states (or phases): solid, liquid, gas, and plasma. However, advances in experimental techniques have revealed other previously theoretical phases, such as Bose–Einstein condensates and fermionic condensates.

Ordinary matter is composed of two types of elementary particles: quarks and leptons. For example, the proton is formed of two up quarks and one down quark; the neutron is formed of two down quarks and one up quark; and the electron is a kind of lepton. An atom consists of an atomic nucleus, made up of protons and neutrons, and electrons that orbit the nucleus. Because most of the mass of an atom is concentrated in its nucleus, which is made up of baryons, astronomers often use the term *baryonic matter* to describe ordinary matter, although a small fraction of this "baryonic matter" is electrons.

Soon after the Big Bang, primordial protons and neutrons formed from the quark–gluon plasma of the early Universe as it cooled below two trillion degrees. A few minutes later, in a process known as Big Bang nucleosynthesis, nuclei formed from the primordial protons and neutrons. This nucleosynthesis formed lighter elements, those with small atomic numbers up to lithium and beryllium, but the abundance of heavier elements dropped off sharply with increasing atomic number. Some boron may have been formed at this time, but the next heavier element, carbon, was not formed in significant amounts. Big Bang nucleosynthesis shut down after about 20 minutes due to the rapid drop in temperature and density of the expanding Universe. Subsequent formation of heavier elements resulted from stellar nucleosynthesis and supernova nucleosynthesis.

Particles

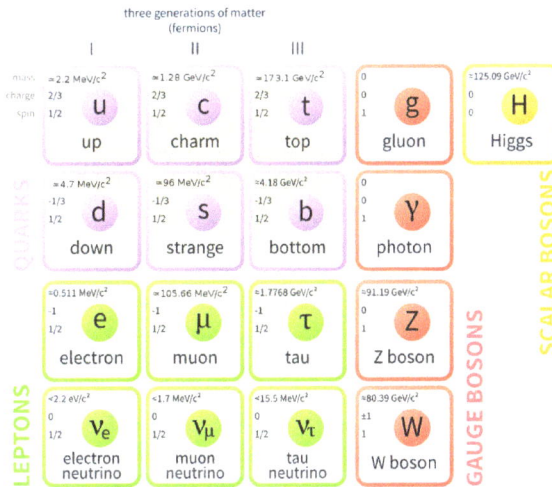

Standard model of elementary particles: the 12 fundamental fermions and 4 fundamental bosons.

Brown loops indicate which bosons (red) couple to which fermions (purple and green). Columns are three generations of matter (fermions) and one of forces (bosons). In the first three columns, two rows contain quarks and two leptons. The top two rows' columns contain up (u) and down (d) quarks, charm (c) and strange (s) quarks, top (t) and bottom (b) quarks, and photon (γ) and gluon (g), respectively. The bottom two rows' columns contain electron neutrino (v_e) and electron (e), muon neutrino (v_μ) and muon (μ), tau neutrino (v_τ) and tau (τ), and the Z^0 and $W^\pm$ carriers of the weak force. Mass, charge, and spin are listed for each particle.

Ordinary matter and the forces that act on matter can be described in terms of elementary particles. These particles are sometimes described as being fundamental, since they have an unknown substructure, and it is unknown whether or not they are composed of smaller and even more fundamental particles. Of central importance is the Standard Model, a theory that is concerned with electromagnetic interactions and the weak and strong nuclear interactions. The Standard Model is supported by the experimental confirmation of the existence of particles that compose matter: quarks and leptons, and their corresponding "antimatter" duals, as well as the force particles that mediate interactions: the photon, the W and Z bosons, and the gluon. The Standard Model predicted the existence of the recently discovered Higgs boson, a particle that is a manifestation of a field within the Universe that can endow particles with mass. Because of its success in explaining a wide variety of experimental results, the Standard Model is sometimes regarded as a "theory of almost everything". The Standard Model does not, however, accommodate gravity. A true force-particle "theory of everything" has not been attained.

Hadrons

A hadron is a composite particle made of quarks held together by the strong force. Hadrons are categorized into two families: baryons (such as protons and neutrons) made of three quarks, and mesons (such as pions) made of one quark and one antiquark. Of the hadrons, protons are stable, and neutrons bound within atomic nuclei are stable. Other hadrons are unstable under ordinary conditions and are thus insignificant constituents of the modern Universe. From approximately 10^{-6} seconds after the Big Bang, during a period is known as the hadron epoch, the temperature of the universe had fallen sufficiently to allow quarks to bind together into hadrons, and the mass of the Universe was dominated by hadrons. Initially the temperature was high enough to allow the formation of hadron/anti-hadron pairs, which kept matter and antimatter in thermal equilibrium. However, as the temperature of the Universe continued to fall, hadron/anti-hadron pairs were no longer produced. Most of the hadrons and anti-hadrons were then eliminated in particle-antiparticle annihilation reactions, leaving a small residual of hadrons by the time the Universe was about one second old.

Leptons

A lepton is an elementary, half-integer spin particle that does not undergo strong interactions but is subject to the Pauli exclusion principle; no two leptons of the same species can be in exactly the same state at the same time. Two main classes of leptons exist: charged leptons (also known as the *electron-like* leptons), and neutral leptons (better known as neutrinos). Electrons are stable and the most common charged lepton in the Universe, whereas muons and taus are unstable particle that quickly decay after being produced in high energy collisions, such as those involving cosmic rays or carried out in particle accelerators. Charged leptons can combine with other particles to form various composite particles such as atoms and positronium. The electron governs nearly all of chemistry, as it is found in atoms and is directly tied to all chemical properties. Neutrinos rarely interact with anything, and are consequently rarely observed. Neutrinos stream throughout the Universe but rarely interact with normal matter.

The lepton epoch was the period in the evolution of the early Universe in which the leptons dominated the mass of the Universe. It started roughly 1 second after the Big Bang, after the majority of hadrons and anti-hadrons annihilated each other at the end of the hadron epoch. During the lepton epoch the temperature of the Universe was still high enough to create lepton/anti-lepton pairs, so leptons and anti-leptons were in thermal equilibrium. Approximately 10 seconds after the Big Bang, the temperature of the Universe had fallen to the point where lepton/anti-lepton pairs were no longer created. Most leptons and anti-leptons were then eliminated in annihilation reactions, leaving a small residue of leptons. The mass of the Universe was then dominated by photons as it entered the following photon epoch.

Photons

A photon is the quantum of light and all other forms of electromagnetic radiation. It is the force carrier for the electromagnetic force, even when static via virtual photons. The effects of this force are easily observable at the microscopic and at the macroscopic level because the photon has zero rest mass; this allows long distance interactions. Like all elementary particles, photons are currently best explained by quantum mechanics and exhibit wave–particle duality, exhibiting properties of waves and of particles.

The photon epoch started after most leptons and anti-leptons were annihilated at the end of the lepton epoch, about 10 seconds after the Big Bang. Atomic nuclei were created in the process of nucleosynthesis which occurred during the first few minutes of the photon epoch. For the remainder of the photon epoch the Universe contained a hot dense plasma of nuclei, electrons and photons. About 380,000 years after the Big Bang, the temperature of the Universe fell to the point where nuclei could combine with electrons to create neutral atoms. As a result, photons no longer interacted frequently with matter and the Universe became transparent. The highly redshifted photons from this period form the cosmic microwave background. Tiny variations in temperature and density detectable in the CMB were the early "seeds" from which all subsequent structure formation took place.

Cosmological Models

Model of the Universe based on General Relativity

General relativity is the geometric theory of gravitation published by Albert Einstein in 1915 and the current description of gravitation in modern physics. It is the basis of current cosmological models of the Universe. General relativity generalizes special relativity and Newton's law of universal gravitation, providing a unified description of gravity as a geometric property of space and time, or spacetime. In particular, the curvature of spacetime is directly related to the energy and momentum of whatever matter and radiation are present. The relation is specified by the Einstein field equations, a system of partial differential equations. In general relativity, the distribution of matter and energy determines the geometry of spacetime, which in turn describes the acceleration of matter. Therefore, solutions of the Einstein field equations describe the evolution of the Universe. Combined with measurements of the amount, type, and distribution of matter in the Universe, the equations of general relativity describe the evolution of the Universe over time.

With the assumption of the cosmological principle that the Universe is homogeneous and isotropic everywhere, a specific solution of the field equations that describes the Universe is the metric tensor called the Friedmann–Lemaître–Robertson–Walker metric,

$$ds^2 = -c^2 dt^2 + R(t)^2 \left(\frac{dr^2}{1-kr^2} + r^2 d\theta^2 + r^2 \sin^2 \theta d\phi^2 \right)$$

where (r, θ, ϕ) correspond to a spherical coordinate system. This metric has only two undetermined parameters. An overall dimensionless length scale factor R describes the size scale of the Universe as a function of time; an increase in R is the expansion of the Universe. A curvature index k describes the geometry. The index k is defined so that it can take only one of three values: 0, corresponding to flat Euclidean geometry; 1, corresponding to a space of positive curvature; or −1, corresponding to a space of positive or negative curvature. The value of R as a function of time t depends upon k and the cosmological constant Λ. The cosmological constant represents the energy density of the vacuum of space and could be related to dark energy. The equation describing how R varies with time is known as the Friedmann equation after its inventor, Alexander Friedmann.

The solutions for *R(t)* depend on k and Λ, but some qualitative features of such solutions are general. First and most importantly, the length scale R of the Universe can remain constant *only* if the Universe is perfectly isotropic with positive curvature (k=1) and has one precise value of density everywhere, as first noted by Albert Einstein. However, this equilibrium is unstable: because the Universe is known to be inhomogeneous on smaller scales, R must change over time. When R changes, all the spatial distances in the Universe change in tandem; there is an overall expansion or contraction of space itself. This accounts for the observation that galaxies appear to be flying apart; the space between them is stretching. The stretching of space also accounts for the apparent paradox that two galaxies can be 40 billion light years apart, although they started from the same point 13.8 billion years ago and never moved faster than the speed of light.

Second, all solutions suggest that there was a gravitational singularity in the past, when R went to zero and matter and energy were infinitely dense. It may seem that this conclusion is uncertain because it is based on the questionable assumptions of perfect homogeneity and isotropy (the cosmological principle) and that only the gravitational interaction is significant. However, the Penrose–Hawking singularity theorems show that a singularity should exist for very general conditions. Hence, according to Einstein's field equations, R grew rapidly from an unimaginably hot, dense state that existed immediately following this singularity (when R had a small, finite value); this is the essence of the Big Bang model of the Universe. Understanding the singularity of the Big Bang likely requires a quantum theory of gravity, which has not yet been formulated.

Third, the curvature index k determines the sign of the mean spatial curvature of spacetime averaged over sufficiently large length scales (greater than about a billion light years). If k=1, the curvature is positive and the Universe has a finite volume. A Universe with positive curvature is often visualized as a three-dimensional sphere embedded in a four-dimensional space. Conversely, if k is zero or negative, the Universe has an infinite volume. It may seem counter-intuitive that an infinite and yet infinitely dense Universe could be created in a single instant at the Big Bang when R=0, but exactly that is predicted mathematically when k does not equal 1. By analogy, an infinite plane has zero curvature but infinite area, whereas an infinite cylinder is finite in one direction and a torus is finite in both. A toroidal Universe could behave like a normal Universe with periodic boundary conditions.

The ultimate fate of the Universe is still unknown, because it depends critically on the curvature index k and the cosmological constant Λ. If the Universe were sufficiently dense, k would equal +1, meaning that its average curvature throughout is positive and the Universe will eventually recollapse in a Big Crunch, possibly starting a new Universe in a Big Bounce. Conversely, if the Universe were insufficiently dense, k would equal 0 or −1 and the Universe would expand forever, cooling off and eventually reaching the Big Freeze and the heat death of the Universe. Modern data suggests that the rate of expansion of the Universe is not decreasing, as originally expected, but increasing; if this continues indefinitely, the Universe may eventually reach a Big Rip. Observationally, the Universe appears to be flat (k = 0), with an overall density that is very close to the critical value between recollapse and eternal expansion.

Multiverse Hypothesis

Some speculative theories have proposed that our Universe is but one of a set of disconnected universes, collectively denoted as the multiverse, challenging or enhancing more limited definitions of the Universe. Scientific multiverse models are distinct from concepts such as alternate planes of consciousness and simulated reality.

Depiction of a multiverse of seven "bubble" universes, which are separate spacetime continua, each having different physical laws, physical constants, and perhaps even different numbers of dimensions or topologies.

Max Tegmark developed a four-part classification scheme for the different types of multiverses that scientists have suggested in response to various Physics problems. An example of such multiverses is the one resulting from the chaotic inflation model of the early universe. Another is the multiverse resulting from the many-worlds interpretation of quantum mechanics. In this interpretation, parallel worlds are generated in a manner similar to quantum superposition and decoherence, with all states of the wave functions being realized in separate worlds. Effectively, in the many-worlds interpretation the multiverse evolves as a universal wavefunction. If the Big Bang that created our multiverse created an ensemble of multiverses, the wave function of the ensemble would be entangled in this sense.

The least controversial category of multiverse in Tegmark's scheme is Level I. The multiverses of this level are composed by distant spacetime events "in our own universe". If space is infinite, or sufficiently large and uniform, identical instances of the history of Earth's entire Hubble volume occur every so often, simply by chance. Tegmark calculated that our nearest so-called doppelgänger, is $10^{10^{115}}$ meters away from us (a double exponential function larger than a googolplex). In principle, it would be impossible to scientifically verify the existence of an identical Hubble volume. However, this existence does follow as a fairly straightforward consequence from otherwise unrelated scientific observations and theories.

It is possible to conceive of disconnected spacetimes, each existing but unable to interact with one another. An easily visualized metaphor of this concept is a group of separate soap bubbles, in which observers living on one soap bubble cannot interact with those on other soap bubbles, even in principle. According to one common terminology, each "soap bubble" of spacetime is denoted as a *universe*, whereas our particular spacetime is denoted as *the Universe*, just as we call our moon *the Moon*. The entire collection of these separate spacetimes is denoted as the multiverse. With this terminology, different *Universes* are not causally connected to each other. In principle, the other unconnected *Universes* may have different dimensionalities and topologies of spacetime, different forms of matter and energy, and different physical laws and physical constants, although such possibilities are purely speculative. Others consider each of several bubbles created as part of chaotic inflation to be separate *Universes*, though in this model these universes all share a causal origin.

Historical Conceptions

Historically, there have been many ideas of the cosmos (cosmologies) and its origin (cosmogonies). Theories of an impersonal Universe governed by physical laws were first proposed by the Greeks and Indians. Ancient Chinese philosophy encompassed the notion of the Universe including both all of space and all of time. Over the centuries, improvements in astronomical observations and theories of motion and gravitation led to ever more accurate descriptions of the Universe. The modern era of cosmology began with Albert Einstein's 1915 general theory of relativity, which made it possible to quantitatively predict the origin, evolution, and conclusion of the Universe as a whole. Most modern, accepted theories of cosmology are based on general relativity and, more specifically, the predicted Big Bang.

Mythologies

Many cultures have stories describing the origin of the world and universe. Cultures generally regard these stories as having some truth. There are however many differing beliefs in how these stories apply amongst those believing in a supernatural origin, ranging from a god directly creating the Universe as it is now to a god just setting the "wheels in motion" (for example via mechanisms such as the big bang and evolution).

Ethnologists and anthropologists who study myths have developed various classification schemes for the various themes that appear in creation stories. For example, in one type of story, the world is born from a world egg; such stories include the Finnish epic poem *Kalevala*, the Chinese story of Pangu or the Indian Brahmanda Purana. In related stories, the Universe is created by a single entity emanating or producing something by him- or herself, as in the Tibetan Buddhism concept of Adi-Buddha, the ancient Greek story of Gaia (Mother Earth), the Aztec goddess Coatlicue myth, the ancient Egyptian god Atum story, and the Judeo-Christian Genesis creation narrative in which the Abrahamic God created the Universe. In another type of story, the Universe is created from the union of male and female deities, as in the Maori story of Rangi and Papa. In other stories, the Universe is created by crafting it from pre-existing materials, such as the corpse of a dead god — as from Tiamat in the Babylonian epic *Enuma Elish* or from the giant Ymir in Norse mythology – or from chaotic materials, as in Izanagi and Izanami in Japanese mythology. In other stories, the Universe emanates from fundamental principles, such as Brahman and Prakrti, the creation myth of the Serers, or the yin and yang of the Tao.

Philosophical Models

The pre-Socratic Greek philosophers and Indian philosophers developed some of the earliest philosophical concepts of the Universe. The earliest Greek philosophers noted that appearances can be deceiving, and sought to understand the underlying reality behind the appearances. In particular, they noted the ability of matter to change forms (e.g., ice to water to steam) and several philosophers proposed that all the physical materials in the world are different forms of a single primordial material, or *arche*. The first to do so was Thales, who proposed this material to be water. Thales' student, Anaximander, proposed that everything came from the limitless *apeiron*. Anaximenes proposed the primordial material to be air on account of its perceived attractive and repulsive qualities that cause the *arche* to condense or dissociate into different forms. Anaxagoras proposed the principle of *Nous* (Mind), while Heraclitus proposed fire (and spoke of *logos*). Empedocles proposed the elements to be earth, water, air and fire. His four-element model became very popular. Like Pythagoras, Plato believed that all things were composed of number, with Empedocles' elements taking the form of the Platonic solids. Democritus, and later philosophers—most notably Leucippus—proposed that the Universe is composed of indivisible atoms moving through a void (vacuum), although Aristotle did not believe that to be feasible because air, like water, offers resistance to motion. Air will immediately rush in to fill a void, and moreover, without resistance, it would do so indefinitely fast.

Although Heraclitus argued for eternal change, his contemporary Parmenides made the radical suggestion that all change is an illusion, that the true underlying reality is eternally unchanging and of a single nature. Parmenides denoted this reality as τὸ ἐν (The One). Parmenides' idea seemed implausible to many Greeks, but his student Zeno of Elea challenged them with several famous paradoxes. Aristotle responded to these paradoxes by developing the notion of a potential countable infinity, as well as the infinitely divisible continuum. Unlike the eternal and unchanging cycles of time, he believed that the world is bounded by the celestial spheres and that cumulative stellar magnitude is only finitely multiplicative.

The Indian philosopher Kanada, founder of the Vaisheshika school, developed a notion of atomism and proposed that light and heat were varieties of the same substance. In the 5th century AD, the Buddhist atomist philosopher Dignāga proposed atoms to be point-sized, durationless, and made of energy. They denied the existence of substantial matter and proposed that movement consisted of momentary flashes of a stream of energy.

The notion of temporal finitism was inspired by the doctrine of creation shared by the three Abrahamic religions: Judaism, Christianity and Islam. The Christian philosopher, John Philoponus, presented the philosophical arguments against the ancient Greek notion of an infinite past and future. Philoponus' arguments against an infinite past were used by the early Muslim philosopher, Al-Kindi (Alkindus); the Jewish philosopher, Saadia Gaon (Saadia ben Joseph); and the Muslim theologian, Al-Ghazali (Algazel).

Astronomical Concepts

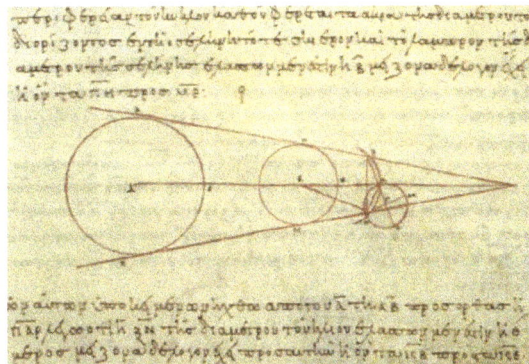

Aristarchus's calculations on the relative sizes of, from left to right, the Sun, Earth, and Moon, from a Greek copy.

Astronomical models of the Universe were proposed soon after astronomy began with the Babylonian astronomers, who viewed the Universe as a flat disk floating in the ocean, and this forms the premise for early Greek maps like those of Anaximander and Hecataeus of Miletus.

Later Greek philosophers, observing the motions of the heavenly bodies, were concerned with developing models of the Universe-based more profoundly on empirical evidence. The first coherent model was proposed by Eudoxus of Cnidos. According to Aristotle's physical interpretation of the model, celestial spheres eternally rotate with uniform motion around a stationary Earth. Normal matter is entirely contained within the terrestrial sphere.

De Mundo (composed before 250 BC or between 350 and 200 BC), stated, "Five elements, situated in spheres in five regions, the less being in each case surrounded by the greater—namely, earth surrounded by water, water by air, air by fire, and fire by ether—make up the whole Universe".

This model was also refined by Callippus and after concentric spheres were abandoned, it was brought into nearly perfect agreement with astronomical observations by Ptolemy. The success of such a model is largely due to the mathematical fact that any function (such as the position of a planet) can be decomposed into a set of circular functions (the Fourier modes). Other Greek scientists, such as the Pythagorean philosopher Philolaus, postulated (according to Stobaeus account) that at the center of the Universe was a "central fire" around which the Earth, Sun, Moon and Planets revolved in uniform circular motion.

The Greek astronomer Aristarchus of Samos was the first known individual to propose a heliocentric model of the Universe. Though the original text has been lost, a reference in Archimedes' book *The Sand Reckoner* describes Aristarchus's heliocentric model. Archimedes wrote:

> You, King Gelon, are aware the Universe is the name given by most astronomers to the sphere the center of which is the center of the Earth, while its radius is equal to the straight line between the center of the Sun and the center of the Earth. This is the common account as you have heard from astronomers. But Aristarchus has brought out a book consisting of certain hypotheses, wherein it appears, as a consequence of the assumptions made, that the Universe is many times greater than the Universe just mentioned. His hypotheses are that the fixed stars and the Sun remain unmoved, that the Earth revolves about the Sun on the circumference of a circle, the Sun lying in the middle of the orbit, and that the sphere of fixed stars, situated about the same center as the Sun, is so great that the circle in which he supposes the Earth to revolve bears such a proportion to the distance of the fixed stars as the center of the sphere bears to its surface

Aristarchus thus believed the stars to be very far away, and saw this as the reason why stellar parallax had not been observed, that is, the stars had not been observed to move relative each other as the Earth moved around the Sun. The stars are in fact much farther away than the distance that was generally assumed in ancient times, which is why stellar parallax is only detectable with precision instruments. The geocentric model, consistent with planetary parallax, was assumed to be an explanation for the unobservability of the parallel phenomenon, stellar parallax. The rejection of the heliocentric view was apparently quite strong, as the following passage from Plutarch suggests (*On the Apparent Face in the Orb of the Moon*):

> Cleanthes [a contemporary of Aristarchus and head of the Stoics] thought it was the duty of the Greeks to indict Aristarchus of Samos on the charge of impiety for putting in motion the Hearth of the Universe [i.e. the Earth],... supposing the heaven to remain at rest and the Earth to revolve in an oblique circle, while it rotates, at the same time, about its own axis

Flammarion engraving, Paris

The only other astronomer from antiquity known by name who supported Aristarchus's helio-centric model was Seleucus of Seleucia, a Hellenistic astronomer who lived a century after Aristarchus. According to Plutarch, Seleucus was the first to prove the heliocentric system through reasoning, but it is not known what arguments he used. Seleucus' arguments for a heliocentric cosmology were probably related to the phenomenon of tides. According to Strabo (1.1.9), Seleucus was the first to state that the tides are due to the attraction of the Moon, and that the height of the tides depends on the Moon's position relative to the Sun. Alternatively, he may have proved helio-centricity by determining the constants of a geometric model for it, and by developing methods to compute planetary positions using this model, like what Nicolaus Copernicus later did in the 16th century. During the Middle Ages, heliocentric models were also proposed by the Indian astronomer Aryabhata, and by the Persian astronomers Albumasar and Al-Sijzi.

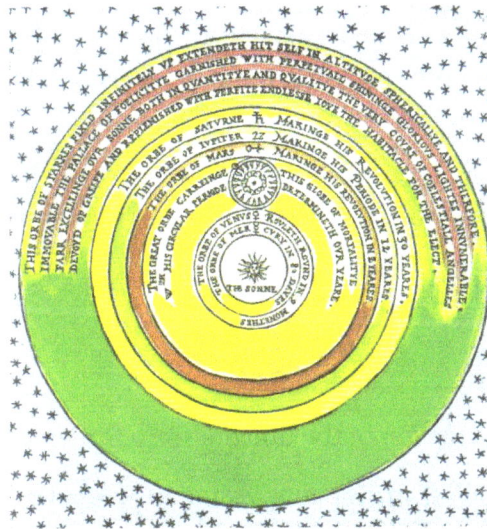

Model of the Copernican Universe with the amendment that the stars are no longer confined to a sphere, but spread uniformly throughout the space surrounding the planets.

The Aristotelian model was accepted in the Western world for roughly two millennia, until Copernicus revived Aristarchus's perspective that the astronomical data could be explained more plausibly if the Earth rotated on its axis and if the Sun were placed at the center of the Universe.

In the center rests the Sun. For who would place this lamp of a very beautiful temple in another or better place than this wherefrom it can illuminate everything at the same time?

—*Nicolaus Copernicus, in Book 1 of De Revolutionibus Orbium Coelestrum (1543)*

As noted by Copernicus himself, the notion that the Earth rotates is very old, dating at least to Philolaus (c. 450 BC), Heraclides Ponticus (c. 350 BC) and Ecphantus the Pythagorean. Roughly a century before Copernicus, the Christian scholar Nicholas of Cusa also proposed that the Earth rotates on its axis in his book, *On Learned Ignorance* (1440). Al-Sijzi also proposed that the Earth rotates on its axis. Empirical evidence for the Earth's rotation on its axis, using the phenomenon of comets, was given by Tusi (1201–1274) and Ali Qushji (1403–1474).

This cosmology was accepted by Isaac Newton, Christiaan Huygens and later scientists. Edmund Halley (1720) and Jean-Philippe de Chéseaux (1744) noted independently that the assumption of an infinite space filled uniformly with stars would lead to the prediction that the nighttime sky would be as bright as the Sun itself; this became known as Olbers' paradox in the 19th century. Newton believed that an infinite space uniformly filled with matter would cause infinite forces and instabilities causing the matter to be crushed inwards under its own gravity. This instability was clarified in 1902 by the Jeans instability criterion. One solution to these paradoxes is the Charlier Universe, in which the matter is arranged hierarchically (systems of orbiting bodies that are themselves orbiting in a larger system, *ad infinitum*) in a fractal way such that the Universe has a negligibly small overall density; such a cosmological model had also been proposed earlier in 1761 by Johann Heinrich Lambert. A significant astronomical advance of the 18th century was the realization by Thomas Wright, Immanuel Kant and others of nebulae.

In 1919, when Hooker Telescope was completed, the prevailing view still was that the Universe consisted entirely of the Milky Way Galaxy. Using the Hooker Telescope, Edwin Hubble identified Cepheid variables in several spiral nebulae and in 1922–1923 proved conclusively that Andromeda Nebula and Triangulum among others, were entire galaxies outside our own, thus proving that Universe consists of multitude of galaxies.

The modern era of physical cosmology began in 1917, when Albert Einstein first applied his general theory of relativity to model the structure and dynamics of the Universe.

Chronology of the Universe

It is convenient to divide the *evolution* of the universe into *three phases*.

In the *first* phase, the very earliest universe is so hot or energetic that initially no matter particles exist, or can exist perhaps only fleetingly. *Space-time itself expands* during an inflationary epoch, due to the immensity of the energies involved. This *inflationary epoch* is the period in the evolution of the early universe when, according to the inflation theory, the universe undergoes an *extremely rapid exponential expansion*. The inflationary epoch lasts from $10-36$ seconds after the Big Bang to sometime between $10-33$ and $10-32$ seconds.

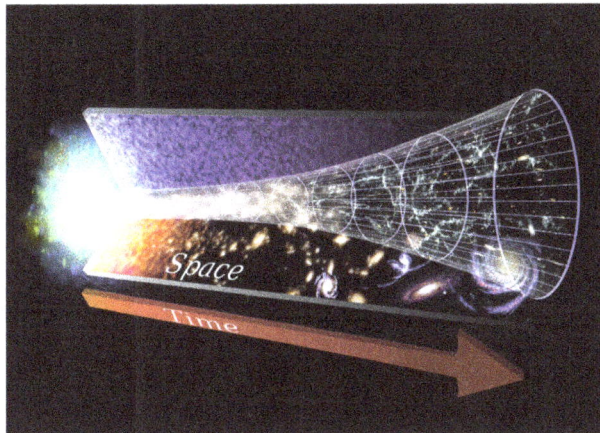

Artist's illustration of the expansion of the Universe.

Gradually the immense energies cool, leading finally to *the first elementary particles of matter* (quarks, gluons, electrons).

In the *second* phase, after the cosmic inflation ended, the early universe is filled with a *quark-gluon plasma*. From this point onwards the physics of the early universe is studied, better understood and less speculative.

For a few *millionths of a second* after the Big Bang, the universe consists of a hot soup of elementary particles, called quarks and gluons. A few microseconds later, those particles begin cooling to form protons and neutrons, the building blocks of matter. Over the past decade, physicists around the world have been trying to re-create that soup, known as quark- gluon plasma (QGP), by slamming together nuclei of atoms with enough energy to produce trillion-degree temperatures.

The *third* phase starts after a 'short' *dark age* (from 300,000 to 150 million years) with a universe whose fundamental particles and forces are as we know them and witnesses the emergence of large scale stable structures, such as *the earliest stars, quasars, galaxies, clusters of galaxies and superclusters* and the development of these to create the kind of universe we see today.

Outline

For the purposes of this summary, it is convenient to divide the chronology of the universe since it originated, into five parts. It is generally considered meaningless or unclear whether time existed before this chronology:

1. The very early universe – the first picosecond (10^{-12}) of cosmic time. It includes the Planck epoch, during which currently understood laws of physics may not apply; the emergence in stages of the four known fundamental interactions or forces – first gravity, and later the strong, weak and electromagnetic interactions; and the expansion of space and supercooling of the still immensely hot universe due to cosmic inflation, which is believed to have been triggered by the separation of the strong and electroweak interaction.

    Tiny ripples in the universe at this stage are believed to be the basis of large-scale structures that formed much later. Different stages of the very early universe are understood to different extents. The earlier parts are beyond the grasp of practical experiments in particle physics but can be explored through other means.

2. The early universe, lasting around 377,000 years. Initially, various kinds of subatomic particles are formed in stages. These particles include almost equal amounts of matter and antimatter, so most of it quickly annihilates, leaving a small excess of matter in the universe.

    At about one second, neutrinos decouple; these neutrinos form the cosmic neutrino background. If primordial black holes exist, they are also formed at about one second of cosmic time. Composite subatomic particles emerge – including protons and neutrons – and from about 3 minutes, conditions are suitable for nucleosynthesis: around 25% of the protons and all the neutrons fuse into heavier elements, mainly helium-4.

    By 20 minutes, the universe is no longer hot enough for fusion, but far too hot for neutral atoms to exist or photons to travel far. It is therefore an opaque plasma. At around 47,000 years, as the universe cools, its behavior begins to be dominated by matter rather than radiation.

    At about 377,000 years, the universe finally becomes cool enough for neutral atoms to form ("recombination"), and as a result it also became transparent for the first time. The newly formed atoms – mainly hydrogen and helium with traces of lithium – quickly reach their lowest energy state (ground state) by releasing photons ("photon decoupling"), and these photons can still be detected today as the cosmic microwave background (CMB). This is currently the oldest observation we have of the universe.

3. Dark Ages and large-scale structure emergence, from 377,000 years until about 1 billion years. After recombination and decoupling, the universe was transparent but stars did not yet exist, so there were no new sources of light (though early in this period many of the existing photons would have had visible-light frequencies, especially red). This period is known as the Dark Ages. The only photons (electromagnetic radiation, or "light") in the universe were the photons released during decoupling (now observed as the cosmic microwave background) and 21 cm wavelength radio emissions occasionally emitted by hydrogen atoms.

    Between about 10 and 17 million years the universe's average temperature was suitable for liquid water (273 – 373K) and there has been speculation whether rocky planets or indeed life could have arisen briefly, since statistically a tiny part of the universe could have had different conditions from the rest, and gained warmth from the universe as a whole.

    At some point around 400 to 700 million years, the earliest generations of stars and galaxies form, and early large structures gradually emerge, drawn to the foam-like dark matter filaments which have already begun to draw together throughout the universe. The earliest generations of stars may have been huge and non-metallic with very short lifetimes compared to most stars we see today, so they commonly finish burning their hydrogen fuel and explode as supernovae after mere millions of years. These early generations of supernovae created most of the everyday elements we see around us today, and seeded the universe with them.

Galaxy clusters and superclusters emerge over time. At some point, high energy photons from the earliest stars, dwarf galaxies and perhaps quasars led to a period of reionization. The universe gradually transitioned into the universe we see around us today, and the Dark Ages only fully came to an end at about 1 billion years.

4. The universe as it appears today. From 1 billion years, and for about 12.8 billions of years, the universe has looked much as it does today. It will continue to appear very similar for many billions of years into the future. The thin disk of our galaxy began to form at about 5 billion years (8.8 bn years ago), and the solar system formed at about 9.2 billion years (4.6 bn years ago), with the earliest traces of life on Earth emerging by about 10.3 billion years (3.5 bn years ago).

From about 9.8 billion years of cosmic time, the slowing expansion of space gradually begins to accelerate under the influence of dark energy, which may be a scalar field throughout our universe. The present-day universe is understood quite well, but beyond about 100 billion years of cosmic time (about 86 billion years in the future), uncertainties in current knowledge mean that we are less sure which path our universe will take.

5. The far future. At some time the Stelliferous Era will end as stars are no longer being born, and the expansion of the universe will mean that the observable universe becomes limited to local galaxies. There are various scenarios for the far future and ultimate fate of the universe. More exact knowledge of our current universe will allow these to be better understood.

In details, earliest stages of chronology shown below (before neutrino decoupling) are an active area of research and based on ideas which are still speculative and subject to modification as scientific knowledge improves.

"Time" column is based on extrapolation of observed metric expansion of space back in the past. For the earliest stages of chronology this extrapolation may be invalid. To give one example, eternal inflation theories propose that inflation lasts forever throughout most of the universe, making the notion of "N seconds since Big Bang" ill-defined.

The radiation temperature refers to the cosmic background radiation and is given by $2.725 \cdot (1+z)$, where z is the redshift.

Epoch	Time	Redshift	Radiation temperature (Energy)	Description
Planck epoch	$<10^{-43}$ s		$>10^{32}$ K ($>10^{19}$ GeV)	The Planck scale is the scale beyond which current physical theories do not have predictive value. The Planck epoch is the time during which physics is assumed to have been dominated by quantum effects of gravity.
Grand unification epoch	$<10^{-36}$ s		$>10^{29}$ K ($>10^{16}$ GeV)	The three forces of the Standard Model are unified (assuming that nature is described by a Grand unification theory).

Inflationary epoch, Electroweak epoch	$<10^{-32}$ s		10^{28} K ~ 10^{22} K (10^{15} ~ 10^9 GeV)	Cosmic inflation expands space by a factor of the order of 10^{26} over a time of the order of 10^{-33} to 10^{-32} seconds. The universe is super-cooled from about 10^{27} down to 10^{22} kelvins. The Strong interaction becomes distinct from the Electroweak interaction.
Quark epoch	10^{-12} s ~ 10^{-6} s		$>10^{12}$ K (>100 MeV)	The forces of the Standard Model have separated, but energies are too high for quarks to coalesce into hadrons, instead forming a quark-gluon plasma. These are the highest energies directly observable in experiment in the Large Hadron Collider.
Hadron epoch	10^{-6} s ~ 1 s		$>10^{10}$ K (>1 MeV)	Quarks are bound into hadrons. A slight matter-antimatter-asymmetry from the earlier phases (baryon asymmetry) results in an elimination of anti-hadrons.
Neutrino decoupling	1 s		10^{10} K (1 MeV)	Neutrinos cease interacting with baryonic matter. The spherical volume of space which will become the Observable universe is approximately 10 light-years in radius at this time.
Lepton epoch	1 s ~ 10 s		10^{10} K ~ 10^9 K (1 MeV ~ 100 keV)	Leptons and anti-leptons remain in thermal equilibrium.
Big Bang nucleosynthesis	10 s ~ 10^3 s		10^9 K ~ 10^7 K (100 keV ~ 1 keV)	Protons and neutrons are bound into primordial atomic nuclei, hydrogen and helium-4. Small amounts of deuterium, helium-3, and lithium-7 are also synthesized. At the end of this epoch, the spherical volume of space which will become the observable universe is about 300 light-years in radius, baryonic matter density is on the order of 4 grams per m³ (about 0.3% of sea level air density) – however, most of energy at this time is in electromagnetic radiation.
Photon epoch	10 s ~ $1.2 \cdot 10^{13}$ s (380 ka)		10^9 K ~ 4000 K (100 keV ~ 0.4 eV)	The universe consists of a plasma of nuclei, electrons and photons; temperatures remain too high for the binding of electrons to nuclei.
Recombination	380 ka	1100	4000 K (0.4 eV)	Electrons and atomic nuclei first become bound to form neutral atoms. Photons are no longer in thermal equilibrium with matter and the Universe first becomes transparent. Recombination lasts for about 100 ka, during which Universe is becoming more and more transparent to photons. The photons of the cosmic microwave background radiation originate at this time. The spherical volume of space which will become the observable universe is 42 million light-years in radius at this time. The baryonic matter density at this time is about 500 million hydrogen and helium atoms per m³, approximately a billion times higher than today. This density corresponds to pressure on the order of 10^{-17} atm.

Dark Ages	380 ka ~ 150 Ma	1100 ~ 20	4000 K ~ 60 K	The time between recombination and the formation of the first stars. During this time, the only source of photons was hydrogen emitting radio waves at hydrogen line. Freely propagating CMB photons quickly (within about 3 million years) red-shifted to infrared, and Universe was devoid of visible light.
Reionization	150 Ma ~ 1 Ga	20 ~ 6	60 K ~ 19 K	The most distant astronomical objects observable with telescopes date to this period; as of 2016, the most remote galaxy observed is GN-z11, at a redshift of 11.09. The earliest "modern" Population III stars are formed in this period.
Galaxy formation and evolution	1 Ga ~ 10 Ga	6 ~ 0.4	19 K ~ 4 K	Galaxies coalesce into "proto-clusters" from about 1 Ga ($z = 6$) and into Galaxy clusters beginning at 3 Gy ($z = 2.1$), and into superclusters from about 5 Gy ($z = 1.2$).
Present time	13.8 Ga	0	2.7 K	Farthest observable photons at this moment are CMB photons. They arrive from a sphere with the radius of 46 billion light-years. The spherical volume inside it is commonly referred to as Observable universe.

Alternative subdivisions of the chronology (overlapping several of the above periods)

Matter-dominated era	47 ka ~ 9.8 Ga	3600 ~ 0.4	10^4 K ~ 4 K	During this time, the energy density of matter dominates both radiation density and dark energy, resulting in a decelerated metric expansion of space.
Dark-energy-dominated era	>9.8 Ga	<0.4	<4 K	Matter density falls below dark energy density (vacuum energy), and expansion of space begins to accelerate. This time happens to correspond roughly to the time of the formation of the Solar System and the evolutionary history of life.
Stelliferous Era	150 Ma ~ 100 Ga	20 ~ −0.99	60 K ~ 0.03 K	The time between the first formation of Population III stars until the cessation of star formation, leaving all stars in the form of degenerate remnants.
Far future	>100 Ga	<−0.99	<0.1 K	The Stelliferous Era will end as stars eventually die and fewer are born to replace them, leading to a darkening universe. Various theories suggest a number of subsequent possibilities. Assuming proton decay, matter may eventually evaporate into a Dark Era (heat death). Alternatively the universe may collapse in a Big Crunch. Alternative suggestions include a false vacuum catastrophe or a Big Rip as possible ends to the universe.

Very Early Universe

Planck Epoch

The Planck epoch is an era in traditional (non-inflationary) big bang cosmology immediately after the event which began our known universe. During this epoch, the temperature and average energies within the universe were so inconceivably high compared to any temperature we can observe today, that everyday subatomic particles could not form, and even the four fundamental forces that shape our universe—electromagnetism, gravitation, weak nuclear interaction, and strong nuclear interaction—were combined and formed one fundamental force. Little is understood about physics at this temperature; different hypotheses propose different scenarios. Traditional big bang cosmology predicts a gravitational singularity before this time, but this theory relies on the theory of general relativity, which is thought to break down for this epoch due to quantum effects.

In inflationary models of cosmology, times before the end of inflation (roughly 10^{-32} second after the Big Bang) do not follow the same timeline as in traditional big bang cosmology. Models that aim to describe the universe and physics during the Planck epoch are generally speculative and fall under the umbrella of "New Physics". Examples include the Hartle–Hawking initial state, string landscape, string gas cosmology, and the ekpyrotic universe.

Grand Unification Epoch

As the universe expanded and cooled, it crossed transition temperatures at which forces separated from each other. These phase transitions can be visualised as similar to condensation and freezing phase transitions of ordinary matter. At certain temperatures/energies, water molecules change their behaviour and structure, and they will behave completely differently. Like steam turning to water, the fields which define our universe's fundamental forces and particles also completely change their behaviors and structures when the temperature/energy falls below a certain point. This is not apparent in everyday life, because it only happens at much, much, higher temperatures than we usually see in our present universe.

These phase transitions are believed to be caused by a phenomenon of quantum fields called "symmetry breaking".

In everyday terms, as the universe cools, it becomes possible for the quantum fields that create the forces and particles around us, to settle at lower energy levels and with higher levels of stability. In doing so, they completely shift how they interact. Forces and interactions arise due to these fields, so the universe can behave very differently above and below a phase transition. For example, in a later epoch, a side effect of one phase transition is that suddenly, many particles that had no mass at all acquire a mass (they begin to interact with the Higgs boson), and a single force begins to manifest as two separate forces.

The grand unification epoch began with a phase transitions of this kind, when gravitation separated from the universal combined gauge force. This caused two forces to now exist: gravity, and an electrostrong interaction. There is no hard evidence yet, that such a combined force existed, but many physicists believe it did. The physics of this electrostrong interaction would be described by a so-called grand unified theory (GUT).

The grand unification epoch ended with a second phase transition, as the electrostrong interaction in turn separated, and began to manifest as two separate interactions, called the strong and electroweak interactions.

Electroweak Epoch

Depending on how epochs are defined, and the model being followed, the electroweak epoch may be considered to start before or after the inflationary epoch. In this article it is described as including the inflationary epoch.

According to traditional big bang cosmology, the electroweak epoch began 10^{-36} seconds after the Big Bang, when the temperature of the universe was low enough (10^{28} K) for the electrostrong interaction to begin to manifest as two separate interactions, called the strong and the electroweak interactions. (The electroweak interaction will also separate later, dividing into the electromagnetic and weak interactions). The exact point where electrostrong symmetry was broken is not certain, because of the very high energies of this event.

In other models of the very early universe, known as inflationary cosmology, the electroweak epoch is said to begin after the inflationary epoch ended, at roughly 10^{-32} seconds.

Inflationary Epoch and the Metric Expansion of Space

At this point, the very early universe suddenly and very rapidly expanded to at least 10^{78} times its previous volume (and possibly much more). This is equivalent to a linear increase of at least 10^{26} times in every spatial dimension – equivalent to an object 1 nanometer (10^{-9} m, about half the width of a molecule of DNA) in length, expanding to one approximately 10.6 light years (about 62 trillion miles) long in a tiny fraction of a second.

Although light and objects within spacetime cannot travel faster than the speed of light, in this case it was the metric governing the size and geometry of spacetime itself that changed in scale. Changes to the metric are not limited by the speed of light.

This change is known as inflation. It is thought to have been triggered by the separation of the strong and electroweak interactions which ended the grand unification epoch. One of the theoretical products of this phase transition was a scalar field called the inflaton field. As this field settled into its lowest energy state throughout the universe, it generated an enormous repulsive force that led to a rapid expansion of space itself. Inflation explains several observed properties of the current universe that are otherwise difficult to account for, including explaining how today's universe has ended up so exceedingly homogeneous (similar) on a very large scale, even though it was highly disordered in its earliest stages.

It is not known exactly when the inflationary epoch ended, but it is thought to have been between 10^{-33} and 10^{-32} seconds after the Big Bang. The rapid expansion of space meant that elementary particles remaining from the grand unification epoch were now distributed very thinly across the universe. However, the huge potential energy of the inflation field was released at the end of the inflationary epoch, as the inflaton field decayed into other particles, known as "reheating". This heating effect led to the universe being repopulated with a dense, hot mixture of quarks, anti-quarks and gluons. Reheating is often considered to mark the start of the electroweak epoch.

In non-traditional versions of Big Bang theory (known as "inflationary" models), inflation ended at a temperature corresponding to roughly 10^{-32} second after the Big Bang, but this does *not* imply that the inflationary era lasted less than 10^{-32} second. To explain the observed homogeneity of the universe, the duration in these models must be longer than 10^{-32} second. Therefore, in inflationary cosmology, the earliest meaningful time "after the Big Bang" is the time of the *end* of inflation.

After inflation ended, the universe continued to expand, but at a very slow rate. The slow expansion began to speed up after several billion years, believed to be due to dark energy, and is still expanding today.

On March 17, 2014, astrophysicists of the BICEP2 collaboration announced the detection of inflationary gravitational waves in the B-mode power spectrum which was interpreted as clear experimental evidence for the theory of inflation. However, on June 19, 2014, lowered confidence in confirming the cosmic inflation findings was reported and finally, on February 2, 2015, a joint analysis of data from BICEP2/Keck and Planck satellite concluded that the statistical "significance [of the data] is too low to be interpreted as a detection of primordial B-modes" and can be attributed mainly to polarized dust in the Milky Way.

Baryogenesis

Baryons are subatomic particles such as protons and neutrons, that are composed of three quarks. It would be expected that both baryons, and particles known as antibaryons would have formed in equal numbers. However, this does not seem to be what happened – as far as we know, the universe was left with far more baryons than antibaryons. Almost no antibaryons are observed in nature. Any explanation for this phenomenon must allow the Sakharov conditions to be satisfied at some time after the end of cosmological inflation. While particle physics suggests asymmetries under which these conditions are met, these asymmetries are too small empirically to account for the observed baryon-antibaryon asymmetry of the universe.

Supersymmetry Breaking (Speculative)

If supersymmetry is a property of our universe, then it must be broken at an energy that is no lower than 1 TeV, the electroweak symmetry scale. The masses of particles and their superpartners would then no longer be equal. This very high energy could explain why no superpartners of known particles have ever been observed.

Electroweak Symmetry Breaking

As the universe's temperature continued to fall below a certain very high energy level (known as the electroweak scale), a third symmetry breaking occurs. So far as we currently know, it was the final symmetry breaking event in the formation of our universe. It is believed that below energies of about 246 GeV, the Higgs field spontaneously acquires a vacuum expectation value. When this happens, it breaks electroweak gauge symmetry. This has two related effects:

1. Via the Higgs mechanism, all elementary particles interacting with the Higgs field become massive, having been massless at higher energy levels.

2. As a side-effect, the weak force and electromagnetic force, and their respective bosons (the W and Z bosons and photon) now begin to manifest differently in the present universe. Before electroweak symmetry breaking these bosons were all massless particles and interacted over long distances, but at this point the W and Z bosons abruptly become massive particles only interacting over distances smaller than the size of an atom, while the photon remains massless and remains a long-distance interaction.

After electroweak symmetry breaking, the fundamental interactions we know of – gravitation, electromagnetism, the strong interaction and the weak interaction – have all taken their present forms, and fundamental particles have mass, but the temperature of the universe is still too high to allow the formation of many fundamental particles we now see in the universe.

Early Universe

After cosmic inflation ends, the universe is filled with a quark–gluon plasma. From this point onwards the physics of the early universe is much better understood, and the energies involved in the Quark epoch are directly amenable to experiment.

The Quark Epoch

The quark epoch began approximately 10^{-12} seconds after the Big Bang. This was the period in the evolution of the early universe immediately after electroweak symmetry breaking, when the fundamental interactions of gravitation, electromagnetism, the strong interaction and the weak interaction had taken their present forms, but the temperature of the universe was still too high to allow quarks to bind together to form hadrons.

During the quark epoch the universe was filled with a dense, hot quark–gluon plasma, containing quarks, leptons and their antiparticles. Collisions between particles were too energetic to allow quarks to combine into mesons or baryons.

The quark epoch ended when the universe was about 10^{-6} seconds old, when the average energy of particle interactions had fallen below the binding energy of hadrons.

Hadron Epoch

The quark–gluon plasma that composes the universe cools until hadrons, including baryons such as protons and neutrons, can form.

Neutrino Decoupling and Cosmic Neutrino Background

At approximately 1 second after the Big Bang neutrinos decouple and begin traveling freely through space. As neutrinos rarely interact with matter, these neutrinos still exist today, analogous to the much later cosmic microwave background emitted during recombination, around 377,000 years after the Big Bang. The neutrinos from this event have a very low energy, around 10^{-10} times smaller than is possible with present-day direct detection. Even high energy neutrinos are notoriously difficult to detect, so this cosmic neutrino background (CNB) may not be directly observed in detail for many years, if at all.

However, Big Bang cosmology makes many predictions about the CNB, and there is very strong indirect evidence that the cosmic neutrino background exists, both from Big Bang nucleosynthesis predictions of the helium abundance, and from anisotropies in the cosmic microwave background. One of these predictions is that neutrinos will have left a subtle imprint on the cosmic microwave background (CMB). It is well known that the CMB has irregularities. Some of the CMB fluctuations were roughly regularly spaced, because of the effect of baryonic acoustic oscillations. In theory, the decoupled neutrinos should have had a very slight effect on the phase of the various CMB fluctuations.

In 2015, it was reported that such shifts had been detected in the CMB. Moreover, the fluctuations corresponded to neutrinos of almost exactly the temperature predicted by Big Bang theory (1.96 +/-0.02K compared to a prediction of 1.95K), and exactly three types of neutrino, the same number of neutrino flavours currently predicted by the Standard Model.

Possible Formation of Primordial Black Holes

Primordial black holes are a hypothetical type of black hole proposed in 1966, that may have formed during the so-called *radiation dominated era*, due to the high densities and inhomogeneous conditions within the first second of cosmic time. Random fluctuations could lead to some regions becoming dense enough to undergo gravitational collapse, forming black holes. Current understandings and theories place tight limits on the abundance and mass of these objects.

Typically, primordial black hole formation requires density contrasts (regional variations in the Universe's density) of around (10%), where is the average density of the Universe. Several mechanisms could produce dense regions meeting this criterion during the early universe, including reheating, cosmological phase transitions and (in so-called "hybrid inflation models") axion inflation. Since primordial black holes didn't form from stellar gravitational collapse, their masses can be far below stellar mass ($\sim 2 \times 10^{33}$ g). Stephen Hawking calculated in 1971 that primordial black holes could weigh as little as 10^{-5} g. But they can have any size, so they could also be large, and may have contributed to the formation of galaxies.

Lepton Epoch

The majority of hadrons and anti-hadrons annihilate each other at the end of the hadron epoch, leaving leptons and anti-leptons dominating the mass of the universe. Approximately 10 seconds after the Big Bang the temperature of the universe falls to the point at which new lepton/anti-lepton pairs are no longer created and most leptons and anti-leptons are eliminated in annihilation reactions, leaving a small residue of leptons.

Photon Epoch

After most leptons and anti-leptons are annihilated at the end of the lepton epoch the energy of the universe is dominated by photons. These photons are still interacting frequently with charged protons, electrons and (eventually) nuclei, and continue to do so for the next 380,000 years.

Nucleosynthesis of Light Elements

Between about 3 and 20 minutes after the Big Bang, the temperature and pressure of the universe allow nuclear fusion to occur, giving rise to nuclei of a few light elements beyond hydrogen ("Big

Bang nucleosynthesis"). About 25% of the protons, and all the neutrons fuse to form deuterium, a hydrogen isotope, and most of the deuterium quickly fuses to form helium-4. The short duration and falling temperature means that only the simplest and fastest fusion processes can occur. Only tiny amounts of nuclei beyond helium are formed, because nucleosynthesis of heavier elements is difficult and requires thousands of years even in stars. Small amounts of tritium (another hydrogen isotope) and beryllium-7 and -8 are formed, but these are unstable and are quickly lost again. A small amount of deuterium is left unfused because of the very short duration.

Therefore, the only stable nuclides created by the end of Big Bang nucleosynthesis are protium (single proton/hydrogen nucleus), deuterium, helium-3, helium-4, and lithium-7. By mass, the resulting matter is about 75% hydrogen nuclei, 25% helium nuclei, and perhaps 10^{-10} by mass of Lithium-7. The next most common stable isotopes produced are lithium-6, beryllium-9, boron-11, carbon, nitrogen and oxygen ("CNO"), but these have predicted abundances of between 5 and 30 parts in 10^{15} by mass, making them essentially undetectable and negligible.

The amounts of each light element in the early universe can be estimated from old galaxies, and is strong evidence for the Big Bang. For example, the Big Bang should produce about 1 neutron for every 7 protons, allowing for 25% of all nucleons to be fused into helium-4 (2 protons and 2 neutrons out of every 16 nucleons), and this is the amount we find today, and far more than can be easily explained by other processes. Similarly, deuterium fuses extremely easily; any alternative explanation must also explain how conditions existed for deuterium to form, but also left some of that deuterium unfused and not immediately fused again into helium. Any alternative must also explain the proportions of the various light elements and their isotopes. A few isotopes, such as lithium-7, were found to be present in amounts that differed from theory, but over time, these differences have been resolved by better observations.

Matter Domination

Until now, the universe's large scale dynamics and behavior have been determined mainly by radiation – meaning, those constituents that move relativistically (at or near the speed of light), such as photons and neutrinos. As the universe cools, from around 47,000 years (z=3600), the universe's large scale behavior becomes dominated by matter instead. This occurs because the energy density of matter begins to exceed both the energy density of radiation and the vacuum energy density. Around or shortly after this time, the densities of non-relativistic matter (atomic nuclei) and relativistic radiation (photons) become equal, the Jeans length, which determines the smallest structures that can form (due to competition between gravitational attraction and pressure effects), begins to fall and perturbations, instead of being wiped out by free-streaming radiation, can begin to grow in amplitude.

According to the Lambda-CDM model, by this stage, the matter in the universe is around 84.5% cold dark matter and 15.5% "ordinary" matter. (However the total matter in the universe is only 31.7%, much smaller than the 68.3% of dark energy). There is overwhelming evidence that dark matter exists and dominates our universe, but since the exact nature of dark matter is still not understood, Big Bang theory does not presently cover any stages in its formation.

From this point on, and for several billion years to come, the presence of dark matter accelerates the formation of structure in our universe. In the early universe, dark matter gradually gathers in huge filaments under the effects of gravity. This amplifies the tiny inhomogeneities (irregularities) in the density of the universe which was left by cosmic inflation. Over time, slightly denser regions become denser and slightly rarefied (emptier) regions become more rarefied. Ordinary matter eventually gathers together faster than it would otherwise do, because of the presence of these concentrations of dark matter.

Recombination, Photon Decoupling, and the Cosmic Microwave Background (CMB)

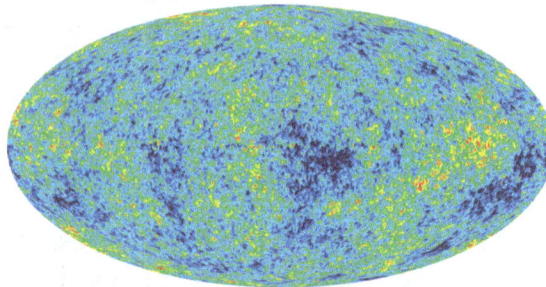

9-year WMAP data shows the cosmic microwave background radiation variations throughout the universe from our perspective, though the actual variations are much smoother than the diagram suggests

About 377,000 years after the Big Bang, two connected events occurred: recombination and photon decoupling. Recombination describes the ionized particles combining to form the first neutral atoms, and decoupling refers to the photons released ("decoupled") as the newly formed atoms settle into more stable energy states.

Just before recombination, the baryonic matter in the universe was at a temperature where it formed a hot ionized plasma. Most of the photons in the universe interacted with electrons and protons, and could not travel significant distances without interacting with ionized particles. As a result, the universe was opaque or "foggy". Although there was light, it was not possible to see, nor can we observe that light through telescopes.

At around 377,000 years, the universe has cooled to a point where free electrons can combine with the hydrogen and helium nuclei to form neutral atoms. This process is relatively fast (and faster for the helium than for the hydrogen), and is known as recombination. The name is slightly inaccurate and is given for historical reasons: in fact the electrons and atomic nuclei were combining for the first time.

Directly combining in a low energy state (ground state) is less efficient, so these hydrogen atoms generally form with the electrons still in a high energy state, and once combined, the electrons quickly release energy in the form of one or more photons as they transition to a low energy state. This release of photons is known as photon decoupling. Some of these decoupled photons are captured by other hydrogen atoms, the remainder remain free. By the end of recombination, most of the protons in the universe have formed neutral atoms. This change from charged to neutral particles means that the mean free path photons can travel before capture in effect becomes infinite, so any decoupled photons that have not been captured can travel freely over long distances. The universe has become transparent to visible light, radio waves and other electromagnetic radiation for the first time in its history.

The photons released by these newly formed hydrogen atoms initially had a temperature/energy of around ~ 4000 K (visible red light). Over billions of years since decoupling, as the universe has expanded, they have red-shifted from visible red light to radio waves (microwave radiation corresponding to a temperature of about 2.7 K). They can still be detected as radio waves today. They form the cosmic microwave background ("CMB"), and they provide crucial evidence of the early universe and how it developed.

Around the same time as recombination, existing pressure waves within the electron-baryon plasma – known as baryon acoustic oscillations – became embedded in the distribution of matter as it condensed, giving rise to a very slight preference in distribution of large-scale objects. Therefore, the cosmic microwave background is a picture of the universe at the end of this epoch including the tiny fluctuations generated during inflation and the spread of objects such as galaxies in the universe is an indication of the scale and size of the universe as it developed over time.

The Dark Ages and Large-scale Structure Emergence

Dark Ages

After recombination and decoupling, the universe was transparent and had cooled enough to allow light to travel long distances, but there were no light-producing structures such as stars and galaxies. Stars and galaxies are formed when dense regions of gas form due to the action of gravity, and this takes a long time within a near-uniform density of gas and on the scale required, so it is estimated that stars did not exist for many millions of years after recombination.

This period, known as the Dark Ages, began around 377,000 years after the Big Bang. During the Dark Ages, the temperature of the universe cooled from some 4000 K down to about 60 K, and only two sources of photons existed: the photons released during recombination/decoupling (as neutral hydrogen atoms formed), which we can still detect today as the cosmic microwave background (CMB), and photons occasionally released by neutral hydrogen atoms, known as the 21 cm spin line of neutral hydrogen.

The October 2010 discovery of UDFy-38135539, the first observed galaxy to have existed during the following reionization epoch, gives us a window into these times. The galaxy earliest in this period observed and thus also the most distant galaxy ever observed is currently on the record of Leiden University's Richard J. Bouwens and Garth D. Illingsworth from UC Observatories/Lick Observatory. They found the galaxy UDFj-39546284 to be at a time some 480 million years after the Big Bang or about halfway through the Cosmic Dark Ages at a distance of about 13.2 billion light-years. More recently, the UDFy-38135539, EGSY8p7 and GN-z11 galaxies were found to be around 380–550 million years after the Big Bang and at a distance of around 13.4 billion light-years. There is also currently an observational effort underway to detect the faint 21 cm spin line radiation, as it is in principle an even more powerful tool than the cosmic microwave background for studying the early universe.

Structures may have begun to emerge from around 150 million years, and stars and early galaxies gradually emerged from around 400 to 700 million years. As they emerged, the Dark Ages gradually ended. Because this process was gradual, the Dark Ages only fully ended around 1 billion (1000 million) years, as the universe took its present appearance.

Habitable Epoch

For about 6.6 million years, between about 10 to 17 million years after the Big Bang (redshift 137–100), the background temperature was between 373 K and 273 K, a temperature compatible with liquid water and common biological chemical reactions. Loeb (2014) speculated that primitive life might in principle have appeared during this window, which he called "the Habitable Epoch of the Early Universe".

At this time, it is usual to say that the only atoms that existed were hydrogen, helium and small traces of other elements, mainly the next heaviest element, lithium. Water is made of hydrogen and oxygen, and all known forms of organic reaction and life require carbon and many other heavier elements than lithium. However it is not precisely correct to say that no other elements were created during the first minutes of the universe – other atoms would have been formed in minuscule quantities. Similarly, there is also a small but non-zero possibility of locally dense concentrations of matter arising, including perhaps densities sufficient for solid planet-sized matter. Loeb therefore speculated that in an amount of matter the size of the universe, extreme statistical anomalies may have created regions where fusion processes by chance had left a concentration of heavier atoms, and discussed whether this might have allowed a window for rocky planets or even life. Warmth would have been available without need for stars such as the sun.

Earliest Structures and Stars Emerge

The Hubble Ultra Deep Fields often showcase galaxies from an ancient era that tell us what the early Stelliferous Age was like.

Another Hubble image shows an infant galaxy forming nearby, which means this happened very recently on the cosmological timescale.

The matter in the universe is around 84.5% cold dark matter and 15.5% "ordinary" matter. Since the start of the matter-dominated era, the dark matter has gradually been gathering in huge spread out (diffuse) filaments under the effects of gravity. Ordinary matter eventually gathers together faster than it would otherwise do, because of the presence of these concentrations of dark matter. It is also slightly more dense at regular distances due to early baryon acoustic oscillations (BAO) which became embedded into the distribution of matter when photons decoupled. Unlike dark matter, ordinary matter can lose energy by many routes, which means that as it collapses, it can lose the energy which would otherwise hold it apart, and collapse more quickly, and into denser forms. Ordinary matter gathers where dark matter is denser, and in those places it collapses into clouds of mainly hydrogen gas. The first stars and galaxies form from these clouds. Where numerous galaxies have formed, galaxy clusters and superclusters will eventually arise. Large voids with few stars will develop between them, marking where dark matter became less common.

Structure formation in the big bang model proceeds hierarchically, due to gravitational collapse, with smaller structures forming before larger ones. The earliest structures to form are the first stars (known as population III stars), dwarf galaxies, and quasars (which are thought to be bright, early active galaxies). Before this epoch, the evolution of the universe could be understood through linear cosmological perturbation theory: that is, all structures could be understood as small deviations from a perfect homogeneous universe. This is computationally relatively easy to study. At this point non-linear structures begin to form, and the computational problem becomes much more difficult, involving, for example, N-body simulations with billions of particles. The Bolshoi Cosmological Simulation is a high precision simulation of this era.

These Population III stars are also responsible for turning the few light elements that were formed in the Big Bang (hydrogen, helium and small amounts of lithium) into many heavier elements. They can be huge as well as perhaps small – and non-metallic (no elements except hydrogen and helium). The larger stars have very short lifetimes compared to most Main Sequence stars we see today, so they commonly finish burning their hydrogen fuel and explode as supernovae after mere millions of years, seeding the universe with heavier elements over repeated generations. They mark the start of the Stelliferous (starry) era.

As yet, no Population III stars have been found, so our understanding of them is based on computational models of their formation and evolution. Fortunately, observations of the Cosmic Microwave Background radiation can be used to date when star formation began in earnest. Analysis of such observations made by the European Space Agency's Planck telescope in 2016 concluded that the first generation of stars formed 700 million years after the Big Bang.

Quasars provides some additional evidence of early structure formation. Their light shows evidence of elements such as carbon, magnesium, iron and oxygen. This is evidence that by the time quasars formed, a massive phase of star formation had already taken place, including sufficient generations of population III stars to give rise to these elements.

Reionization

As the first stars, dwarf galaxies and quasars gradually form, the intense radiation they emit reionizes much of the surrounding universe; splitting the neutral hydrogen atoms back into a plasma of free electrons and protons for the first time since recombination and decoupling.

Reionization is evidenced from observations of quasars. Quasars are a form of active galaxy, and the most luminous objects observed in the universe. Electrons in neutral hydrogen have a specific patterns of absorbing photons, related to electron energy levels and called the Lyman series. Ionized hydrogen does not have electron energy levels of this kind. Therefore, light travelling through ionized hydrogen and neutral hydrogen shows different absorption lines. In addition, the light will have travelled for billions of years to reach us, so any absorption by neutral hydrogen will have been redshifted by varied amounts, rather than by one specific amount, indicating when it happened. These features make it possible to study the state of ionization at many different times in the past. They show that reionization began as "bubbles" of ionized hydrogen which became larger over time. They also show that the absorption was due to the general state of the universe (the intergalactic medium) and not due to passing through galaxies or other dense areas. Reionization might have started as early as $z=16$ (250 million years of cosmic time) and was complete by around $z=9$ or 10 (500 million years). The epoch of reionization probably ended by around $z=5$ or 6 (1 billion years) as the era of Population III stars and quasars – and their intense radiation – came to an end, and the ionized hydrogen gradually reverted to neutral atoms.

These observations have narrowed down the period of time during which reionization took place, but the source of the photons that caused reionization is still not completely certain. To ionize neutral hydrogen, an energy larger than 13.6 eV is required, which corresponds to ultraviolet photons with a wavelength of 91.2 nm or shorter, implying that the sources must have produced significant amount of ultraviolet and higher energy. Protons and electrons will recombine if energy is not continuously provided to keep them apart, which also sets limits on how numerous the sources where and their longevity. With these constraints, it is expected that quasars and first generation stars and galaxies were the main sources of energy. The current leading candidates from most to least significant are currently believed to be population III stars (the earliest stars) (possibly 70%), dwarf galaxies (very early small high-energy galaxies) (possibly 30%), and a contribution from quasars (a class of active galactic nuclei).

However, by this time, matter had become far more spread out due to the ongoing expansion of the universe. Although the neutral hydrogen atoms were again ionized, the plasma was much more thin and diffuse, and photons were much less likely to be scattered. Despite being reionized, the universe remained largely transparent during reionization. As the universe continued to cool and expand, reionization gradually ended.

Galaxies, Clusters and Superclusters

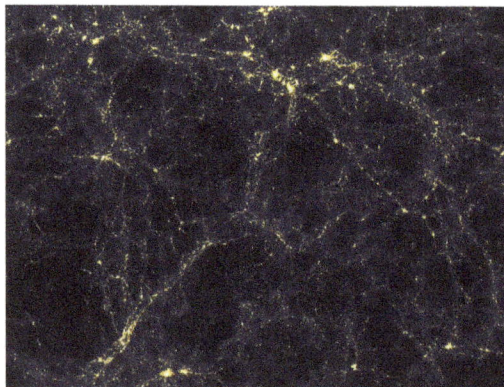

Computer simulated view of the large-scale structure of a part of the
universe about 50 million light years across

Matter continues to draw together under the influence of gravity, to form galaxies. The stars from this time period, known as Population II stars, are formed early on in this process, with Population I stars formed later. Gravitational attraction also gradually pulls galaxies towards each other to form groups, clusters and superclusters. The Hubble Ultra Deep Field observatory has identified a number of small galaxies merging to form larger ones, at 800 million years of cosmic time (13 billion years ago) (this age estimate is now believed to be slightly overstated).

Johannes Schedler's project has identified a quasar CFHQS 1641+3755 at 12.7 billion light-years away, when the universe was just 7% of its present age. On July 11, 2007, using the 10-metre Keck II telescope on Mauna Kea, Richard Ellis of the California Institute of Technology at Pasadena and his team found six star forming galaxies about 13.2 billion light years away and therefore created when the universe was only 500 million years old. Only about 10 of these extremely early objects are currently known. More recent observations have shown these ages to be shorter than previously indicated. The most distant galaxy observed as of October 2016, GN-z11, has been reported to be 32 billion light years away, a vast distance made possible through space-time expansion (redshift z=11.1; comoving distance of 32 billion light-years; lookback time of 13.4 billion years).

Current Appearance of the Universe

The universe has appeared much the same as it does now, for many billions of years. It will continue to look similar for many more billions of years into the future.

Based upon the emerging science of nucleocosmochronology, the Galactic thin disk of the Milky Way is estimated to have been formed 8.8 ± 1.7 billion years ago.

Dark Energy Dominated Era

From about 9.8 billion years of cosmic time, the universe's large-scale behavior is believed to have gradually changed for a third time. Previously it had been dominated by radiation (relativistic constituents) for the first 47,000 years, and then by matter. From this point, the expansion of the universe gradually begins to accelerate. Observation confirms the increasing expansion. While the precise cause is not known, by far the most accepted understanding is that this is due to an unknown form of energy called dark energy. Research is ongoing to understand this dark energy, which is believed to constitute about 68.3% of the entire physical universe.

Dark energy acts like a cosmological constant - a scalar field that exists throughout space. Therefore, unlike gravity, its effects do not diminish (or only diminish slowly) as the universe grows. By contrast, matter and gravity have a greater effect initially, but it diminishes quicker as the universe continues to expand. Over time, the deceleration and inward attraction due to gravity reduces more quickly. Eventually the outward and repulsive effect of dark energy increasingly dominates, and the expansion of space starts to slowly accelerate.

Far Future and Ultimate Fate

The universe has existed for around 13.8 billion years, and we believe that we understand it well enough to predict its large-scale development for many billions of years into the future – perhaps as much as 100 billion years of cosmic time (about 86 billion years from now). Beyond that, we need better understandings, to make accurate predictions. Therefore, the universe could follow a variety of different paths beyond this time.

There are several competing scenarios for the possible long-term evolution of the universe. Which of them will happen, if any, depends on the precise values of physical constants such as the cosmological constant, the possibility of proton decay, and the natural laws beyond the Standard Model.

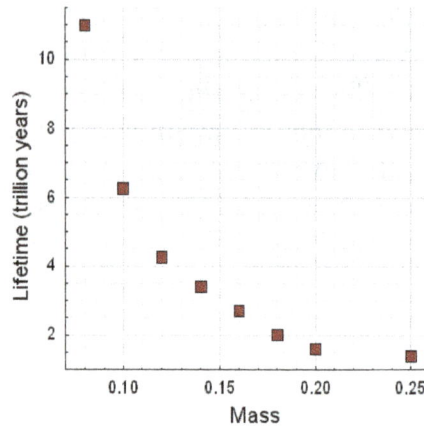

The predicted main-sequence lifetime of a red dwarf plotted against its mass relative to the Sun

If the expansion of the universe continues and it stays in its present form, eventually all but the nearest galaxies will be carried away from us by the expansion of space at such a velocity that our observable universe will be limited to our own gravitationally bound local galactic cluster. In the very long term (after many trillions – thousands of billions – of years, cosmic time), the Stelliferous Era will end, as stars cease to be born and even the longest-lived stars gradually die. Beyond this, all objects in the universe will cool and (with the possible exception of protons) gradually decompose back to their constituent particles and then into subatomic particles and very low level photons and other fundamental particles, by a variety of possible processes. But this will take a duration of time that is almost inconceivable to most people, compared to which the entire 13.8 billion years of the universe would be a tiny instant in time.

Ultimately, in the extreme future, the following scenarios have been proposed for the ultimate fate of the universe.

| **Heat Death** | As expansion continues, the universe becomes larger, colder, and more dilute; in time, all structures eventually decompose to subatomic particles and photons. | In the case of indefinitely continuing metric expansion of space, the energy density in the universe will decrease until, after an estimated time of 10^{1000} years, it reaches thermodynamic equilibrium and no more structure will be possible. This will happen only after an extremely long time because first, all matter will collapse into black holes, which will then evaporate extremely slowly via Hawking radiation. The universe in this scenario will cease to be able to support life much earlier than this, after some 10^{14} years or so, when star formation ceases.[§IID.] In some grand unified theories, proton decay after at least 10^{34} years will convert the remaining interstellar gas and stellar remnants into leptons (such as positrons and electrons) and photons. Some positrons and electrons will then recombine into photons.[§IV, §VF.] In this case, the universe has reached a high-entropy state consisting of a bath of particles and low-energy radiation. It is not known however whether it eventually achieves thermodynamic equilibrium.[§VIB, VID.] The hypothesis of a universal heat death stems from the 1850s ideas of William Thomson (Lord Kelvin) who extrapolated the theory of heat views of mechanical energy loss in nature, as embodied in the first two laws of thermodynamics, to universal operation. |

Big Rip	Expansion of space accelerates and at some point becomes so extreme that even subatomic particles and the fabric of spacetime are pulled apart and unable to exist.	For sufficiently large values for the dark energy content of the universe, the expansion rate of the universe will continue to increase without limit. Gravitationally bound systems, such as clusters of galaxies, galaxies, and ultimately the Solar System will be torn apart. Eventually the expansion will be so rapid as to overcome the electromagnetic forces holding molecules and atoms together. Even atomic nuclei will be torn apart. Finally, forces and interactions even on the Planck scale – the smallest size for which the notion of "space" currently has a meaning – will no longer be able to occur as the fabric of spacetime itself is pulled apart and the universe as we know it will end in an unusual kind of singularity.
Big Crunch	Expansion eventually slows and halts, then reverses as all matter accelerates towards its common centre. Not now considered likely.	In the opposite of the "Big Rip" scenario, the metric expansion of space would at some point be reversed and the universe would contract towards a hot, dense state. This is a required element of oscillatory universe scenarios, such as the cyclic model, although a Big Crunch does not necessarily imply an oscillatory universe. Current observations suggest that this model of the universe is unlikely to be correct, and the expansion will continue or even accelerate.
Vacuum instability	Collapse of the quantum fields that underpin all forces, particles and structures, to a different form.	Cosmology traditionally has assumed a stable or at least metastable universe, but the possibility of a false vacuum in quantum field theory implies that the universe at any point in spacetime might spontaneously collapse into a lower energy state, a more stable or "true vacuum", which would then expand outward from that point with the speed of light.
		The effect would be that the quantum fields that underpin all forces, particles and structures, would undergo a transition to a more stable form. New forces and particles would replace the present ones we know of, with the side effect that all current particles, forces and structures would be destroyed and subsequently (if able) reform into different particles, forces and structures.

In this kind of extreme timescale, extremely rare quantum phenomenae may also occur that are extremely unlikely to be seen on a timescale smaller than trillions of years. These may also lead to unpredictable changes to the state of the universe which would not be likely to be significant on any smaller timescale. For example, on a timescale of millions of trillions of years, black holes might appear to evaporate almost instantly, uncommon quantum tunneling phenomenae would appear to be common, and quantum (or other) phenomenae so unlikely that they might occur just once in a trillion years may occur many, many times.

Observable Universe

The diameter of the observable universe is estimated at about 28 billion parsecs (93 billion light-years). As a reminder, a light-year is a unit of length equal to just under 10 trillion kilometres (or about 6 trillion miles).

The Observable Universe consists of the galaxies and other matter that we can, in principle, observe from Earth in the present day—because light (or other signals) from those objects has had time to reach the Earth since the beginning of the cosmological expansion.

The numbers are pretty hard to comprehend even when you know what each unit represents. To even think of how long 10 trillion kilometers might be, let alone 93 billion times that distance, can cause your brain to hurt. Andrew Z. Colvin has attempted to put some of this incomprehensible size into perspective by starting with our own planet and zooming out from there.

Size of the Observable Universe

We can only observe the universe by looking at particles, e.g. photons, i.e. light, that reaches us. Since locally nothing can travel faster than the speed of light, the distance that light could have travelled within the age of the universe, t o determines the size of the observable universe.

The distance to the particle horizon, r_p, is the distance of an object that has emitted particles (light), which reach us today, and were emitted at $t = 0$, i.e. one age of the universe ago.

Which kind of distance? Proper distance, $D := Rx$, or comoving distance,

A reasonable thing would be to ask for the size of the universe as it is today, i.e. to ask for the proper distance, as shown in the figure.

But, most commonly, the scale factor $R(t)$ is defined such that $R(t_0) = 1$. Therefore, if we ask for today's particle horizon, $r_p(t = t_0)$, there is no difference.

$$r_p := D_p(t_0) = x_p R(t_0), R(t_0) = 1$$

Misconceptions on its Size

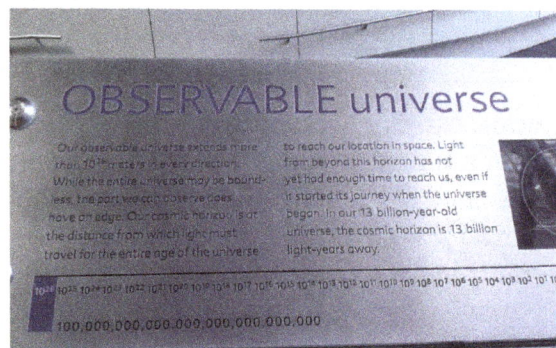

An example of the misconception that the radius of the observable universe is 13 billion light-years.

Many secondary sources have reported a wide variety of incorrect figures for the size of the visible universe. Some of these figures are listed below, with brief descriptions of possible reasons for misconceptions about them.

13.8 billion light-years

> The age of the universe is estimated to be 13.8 billion years. While it is commonly understood that nothing can accelerate to velocities equal to or greater than that of light, it is a common misconception that the radius of the observable universe must therefore amount to only 13.8 billion light-years. This reasoning would only make sense if the flat, static Minkowski spacetime conception under special relativity were correct. In the real universe,

spacetime is curved in a way that corresponds to the expansion of space, as evidenced by Hubble's law. Distances obtained as the speed of light multiplied by a cosmological time interval have no direct physical significance.

15.8 billion light-years

This is obtained in the same way as the 13.8-billion-light-year figure, but starting from an incorrect age of the universe that the popular press reported in mid-2006.

27.6 billion light-years

This is a diameter obtained from the (incorrect) radius of 13.8 billion light-years.

78 billion light-years

In 2003, Cornish *et al.* found this lower bound for the diameter of the whole universe (not just the observable part), postulating that the universe is finite in size due to its having a nontrivial topology, with this lower bound based on the estimated current distance between points that we can see on opposite sides of the cosmic microwave background radiation (CMBR). If the whole universe is smaller than this sphere, then light has had time to circumnavigate it since the Big Bang, producing multiple images of distant points in the CMBR, which would show up as patterns of repeating circles. Cornish et al. looked for such an effect at scales of up to 24 gigaparsecs (78 Gly or 7.4×10^{26} m) and failed to find it, and suggested that if they could extend their search to all possible orientations, they would then "be able to exclude the possibility that we live in a universe smaller than 24 Gpc in diameter". The authors also estimated that with "lower noise and higher resolution CMB maps (from WMAP's extended mission and from Planck), we will be able to search for smaller circles and extend the limit to ~28 Gpc." This estimate of the maximum lower bound that can be established by future observations corresponds to a radius of 14 gigaparsecs, or around 46 billion light-years, about the same as the figure for the radius of the visible universe (whose radius is defined by the CMBR sphere) given in the opening section. A 2012 preprint by most of the same authors as the Cornish et al. paper has extended the current lower bound to a diameter of 98.5% the diameter of the CMBR sphere, or about 26 Gpc.

156 billion light-years

This figure was obtained by doubling 78 billion light-years on the assumption that it is a radius. Because 78 billion light-years is already a diameter (the original paper by Cornish et al. says, "By extending the search to all possible orientations, we will be able to exclude the possibility that we live in a universe smaller than 24 Gpc in diameter," and 24 Gpc is 78 billion light-years), the doubled figure is incorrect. This figure was very widely reported. A press release from Montana State University–Bozeman, where Cornish works as an astrophysicist, noted the error when discussing a story that had appeared in *Discover* magazine, saying "*Discover* mistakenly reported that the universe was 156 billion light-years wide, thinking that 78 billion was the radius of the universe instead of its diameter." As noted above, 78 billion was also incorrect.

180 billion light-years

> This estimate combines the erroneous 156-billion-light-year figure with evidence that the M33 Galaxy is actually fifteen percent farther away than previous estimates and that, therefore, the Hubble constant is fifteen percent smaller. The 180-billion figure is obtained by adding 15% to 156 billion light-years.

Large-scale Structure

Sky surveys and mappings of the various wavelength bands of electromagnetic radiation (in particular 21-cm emission) have yielded much information on the content and character of the universe's structure. The organization of structure appears to follow as a hierarchical model with organization up to the scale of superclusters and filaments. Larger than this (at scales between 30 and 200 megaparsecs), there seems to be no continued structure, a phenomenon that has been referred to as the *End of Greatness*.

Walls, Filaments, Nodes, and Voids

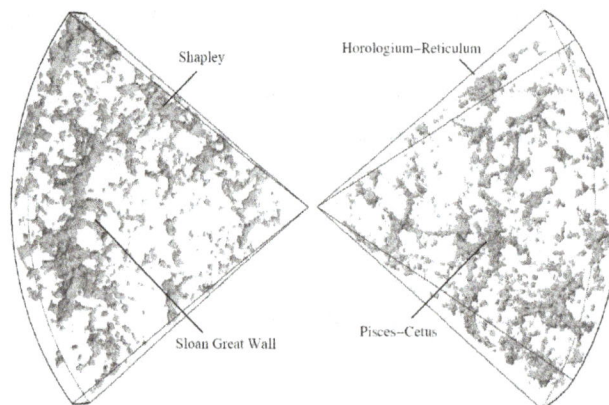

DTFE reconstruction of the inner parts of the 2dF Galaxy Redshift Survey

The organization of structure arguably begins at the stellar level, though most cosmologists rarely address astrophysics on that scale. Stars are organized into galaxies, which in turn form galaxy groups, galaxy clusters, superclusters, sheets, walls and filaments, which are separated by immense voids, creating a vast foam-like structure sometimes called the "cosmic web". Prior to 1989, it was commonly assumed that virialized galaxy clusters were the largest structures in existence, and that they were distributed more or less uniformly throughout the universe in every direction. However, since the early 1980s, more and more structures have been discovered. In 1983, Adrian Webster identified the Webster LQG, a large quasar group consisting of 5 quasars. The discovery was the first identification of a large-scale structure, and has expanded the information about the known grouping of matter in the universe. In 1987, Robert Brent Tully identified the Pisces–Cetus Supercluster Complex, the galaxy filament in which the Milky Way resides. It is about 1 billion light-years across. That same year, an unusually large region with no galaxies was discovered, the Giant Void, which measures 1.3 billion light-years across. Based on redshift survey data, in 1989 Margaret Geller and John Huchra discovered the "Great Wall", a sheet of galaxies more than 500 million light-years long and 200 million light-years wide, but only 15 million light-years thick. The existence of this structure escaped notice for so long because it requires locating the position of galaxies in three dimensions, which involves combining location information about the galaxies with distance information from redshifts. Two years later, astronomers Roger G. Clowes and Luis E. Campusano discovered the Clowes–Campusano LQG, a large quasar group measuring two billion light-years at its widest point which was the largest known structure in the universe at the time of its announcement. In April 2003, another large-scale structure was discovered, the Sloan Great Wall. In August 2007, a possible supervoid was detected in the constellation Eridanus. It coincides with the 'CMB cold spot', a cold region in the microwave sky that is highly improbable under the currently favored cosmological model. This supervoid could cause the cold spot, but to do so it would have to be improbably big, possibly a billion light-years across, almost as big as the Giant Void mentioned above.

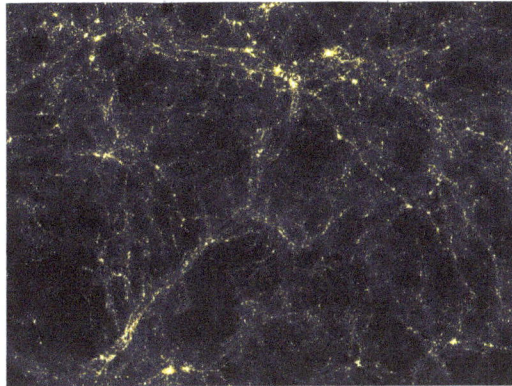

Computer simulated image of an area of space more than 50 million light-years across, presenting a possible large-scale distribution of light sources in the universe—precise relative contributions of galaxies and quasars are unclear.

Another large-scale structure is the Newfound Blob, a collection of galaxies and enormous gas bubbles that measures about 200 million light-years across.

In 2011, a large quasar group was discovered, U1.11, measuring about 2.5 billion light-years across. On January 11, 2013, another large quasar group, the Huge-LQG, was discovered, which was measured to be four billion light-years across, the largest known structure in the universe at that time. In November 2013, astronomers discovered the Hercules–Corona Borealis Great Wall, an even bigger structure twice as large as the former. It was defined by the mapping of gamma-ray bursts.

End of Greatness

The *End of Greatness* is an observational scale discovered at roughly 100 Mpc (roughly 300 million light-years) where the lumpiness seen in the large-scale structure of the universe is homogenized and isotropized in accordance with the Cosmological Principle. At this scale, no pseudo-random fractalness is apparent. The superclusters and filaments seen in smaller surveys are randomized to the extent that the smooth distribution of the universe is visually apparent. It was not until the redshift surveys of the 1990s were completed that this scale could accurately be observed.

Observations

Another indicator of large-scale structure is the 'Lyman-alpha forest'. This is a collection of absorption lines that appear in the spectra of light from quasars, which are interpreted as indicating the existence of huge thin sheets of intergalactic (mostly hydrogen) gas. These sheets appear to be associated with the formation of new galaxies.

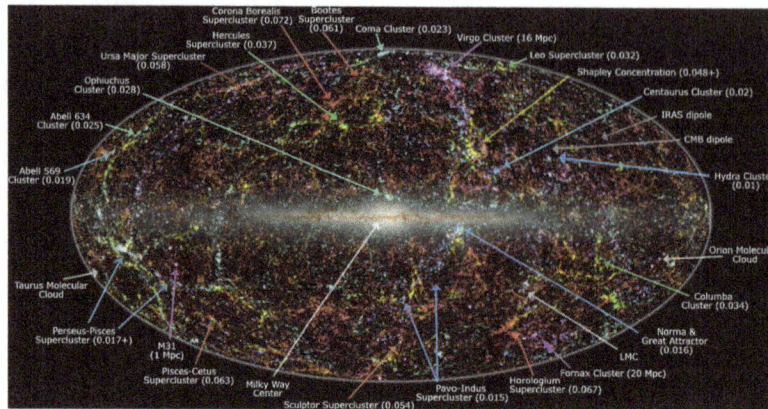

Panoramic view of the entire near-infrared sky reveals the distribution of galaxies beyond the Milky Way.

The galaxies are color-coded by 'redshift' obtained from the UGC, CfA, Tully NBGC, LCRS, 2dF, 6dFGS, and SDSS surveys (and from various observations compiled by the NASA Extragalactic Database), or photo-metrically deduced from the K band (2.2 µm). Blue are the nearest sources ($z < 0.01$); green are at moderate distances ($0.01 < z < 0.04$) and red are the most distant sources that 2MASS resolves ($0.04 < z < 0.1$). The map is projected with an equal area Aitoff in the Galactic system (Milky Way at center).

Caution is required in describing structures on a cosmic scale because things are often different from how they appear. Gravitational lensing (bending of light by gravitation) can make an image appear to originate in a different direction from its real source. This is caused when foreground objects (such as galaxies) curve surrounding spacetime (as predicted by general relativity), and deflect passing light rays. Rather usefully, strong gravitational lensing can sometimes magnify distant galaxies, making them easier to detect. Weak lensing (gravitational shear) by the intervening universe in general also subtly changes the observed large-scale structure.

The large-scale structure of the universe also looks different if one only uses redshift to measure distances to galaxies. For example, galaxies behind a galaxy cluster are attracted to it, and so fall towards it, and so are slightly blueshifted (compared to how they would be if there were no cluster) On the near side, things are slightly redshifted. Thus, the environment of the cluster looks a bit squashed if using redshifts to measure distance. An opposite effect works on the galaxies already within a cluster: the galaxies have some random motion around the cluster center, and when these random motions are converted to redshifts, the cluster appears elongated. This creates a *"finger of God"*—the illusion of a long chain of galaxies pointed at the Earth.

Cosmography of Earth's Cosmic Neighborhood

At the centre of the Hydra-Centaurus Supercluster, a gravitational anomaly called the Great Attractor affects the motion of galaxies over a region hundreds of millions of light-years across. These galaxies are all redshifted, in accordance with Hubble's law. This indicates that they are receding from us and from each other, but the variations in their redshift are sufficient to reveal the existence of a concentration of mass equivalent to tens of thousands of galaxies.

The Great Attractor, discovered in 1986, lies at a distance of between 150 million and 250 million light-years (250 million is the most recent estimate), in the direction of the Hydra and Centaurus constellations. In its vicinity there is a preponderance of large old galaxies, many of which are colliding with their neighbours, or radiating large amounts of radio waves.

In 1987, astronomer R. Brent Tully of the University of Hawaii's Institute of Astronomy identified what he called the Pisces–Cetus Supercluster Complex, a structure one billion light-years long and 150 million light-years across in which, he claimed, the Local Supercluster was embedded.

Mass of Ordinary Matter

The mass of the observable universe is often quoted as 10^{50} tonnes or 10^{53} kg. In this context, mass refers to ordinary matter and includes the interstellar medium (ISM) and the intergalactic medium (IGM). However, it excludes dark matter and dark energy. This quoted value for the mass of ordinary matter in the universe can be estimated based on critical density. The calculations are for the observable universe only as the volume of the whole is unknown and may be infinite.

Estimates Based on Critical Density

Critical density is the energy density for which the universe is flat. If there is no dark energy, it is also the density for which the expansion of the universe is poised between continued expansion and collapse. From the Friedmann equations, the value for critical density, is:

$$\rho_c = \frac{3H^2}{8\pi G},$$

where G is the gravitational constant and H_o is the present value of the Hubble constant. The current value for H_o, due to the European Space Agency's Planck Telescope, is H_o = 67.15 kilometers per second per mega parsec. This gives a critical density of 0.85×10^{-26} kg/m³ (commonly quoted as about 5 hydrogen atoms per cubic meter). This density includes four significant types of energy/mass: ordinary matter (4.8%), neutrinos (0.1%), cold dark matter (26.8%), and dark energy (68.3%). Note that although neutrinos are defined as particles like electrons, they are listed separately because they are difficult to detect and so different from ordinary matter. The density of ordinary matter, as measured by Planck, is 4.8% of the total critical density or 4.08×10^{-28} kg/m³. To convert this density to mass we must multiply by volume, a value based on the radius of the "observable universe". Since the universe has been expanding for 13.8 billion years, the comoving distance (radius) is now about 46.6 billion light-years. Thus, volume $(4/3\pi r^3)$ equals 3.58×10^{80} m³ and the mass of ordinary matter equals density $(4.08\times10^{-28}$ kg/m³) times volume $(3.58\times10^{80}$ m³) or 1.46×10^{53} kg.

Matter Content – Number of Atoms

Assuming the mass of ordinary matter is about 1.45×10^{53} kg (refer to previous section) and assuming all atoms are hydrogen atoms (which in reality make up about 74% of all atoms in our galaxy by mass), calculating the estimated total number of atoms in the observable universe is straightforward. Divide the mass of ordinary matter by the mass of a hydrogen atom (1.45×10^{53} kg divided by 1.67×10^{-27} kg). The result is approximately 10^{80} hydrogen atoms.

Most Distant Objects

The most distant astronomical object yet announced as of January 2011 is a galaxy candidate classified UDFj-39546284. In 2009, a gamma ray burst, GRB 090423, was found to have a redshift of 8.2, which indicates that the collapsing star that caused it exploded when the universe was only 630 million years old. The burst happened approximately 13 billion years ago, so a distance of about 13 billion light-years was widely quoted in the media (or sometimes a more precise figure of 13.035 billion light-years), though this would be the "light travel distance" rather than the "proper distance" used in both Hubble's law and in defining the size of the observable universe (cosmologist Ned Wright argues against the common use of light travel distance in astronomical press releases on this page, and at the bottom of the page offers online calculators that can be used to calculate the current proper distance to a distant object in a flat universe based on either the redshift z or the light travel time). The proper distance for a redshift of 8.2 would be about 9.2 Gpc, or about 30 billion light-years. Another record-holder for most distant object is a galaxy observed through and located beyond Abell 2218, also with a light travel distance of approximately 13 billion light-years from Earth, with observations from the Hubble telescope indicating a redshift between 6.6 and 7.1, and observations from Keck telescopes indicating a redshift towards the upper end of this range, around 7. The galaxy's light now observable on Earth would have begun to emanate from its source about 750 million years after the Big Bang.

Horizons

The limit of observability in our universe is set by a set of cosmological horizons which limit—based on various physical constraints—the extent to which we can obtain information about various events in the universe. The most famous horizon is the particle horizon which sets a limit on the precise distance that can be seen due to the finite age of the universe. Additional horizons are associated with the possible future extent of observations (larger than the particle horizon owing to the expansion of space), an "optical horizon" at the surface of last scattering, and associated horizons with the surface of last scattering for neutrinos and gravitational waves.

A diagram of our location in the observable universe.

Galaxy Cluster

Cluster of galaxies, Gravitationally bound grouping of galaxies, numbering from the hundreds to the tens of thousands. Large clusters of galaxies often exhibit extensive X-ray emission from intergalactic gas heated to tens of millions of degrees. Also, interactions of galaxies with each other and with the intracluster gas may deplete galaxies of their own interstellar gas. The Milky Way Galaxy belongs to the Local Group, which lies on the outskirts of the Virgo Cluster

Formation of Galaxy

Formation of galaxy clusters corresponds to the collapse of the largest gravitationally bound overdensities in the initial density field and is accompanied by the most energetic phenomena since the Big

Bang and by the complex interplay between gravity-induced dynamics of collapse and baryonic processes associated with galaxy formation. Galaxy clusters are, thus, at the cross-roads of cosmology and astrophysics and are unique laboratories for testing models of gravitational structure formation, galaxy evolution, thermodynamics of the intergalactic medium, and plasma physics. At the same time, their large masses make them a useful probe of growth of structure over cosmological time, thus providing cosmological constraints that are complementary to other probes.

Age of the Universe

The age of the universe is approximately 13.77 billion years. This age is calculated by measuring the distances and radial velocities of other galaxies, most of which are flying away from our own at speeds proportional to their distances. Using the current expansion rate of the universe, we can imagine "rewinding" the universe to the point where everything was contained in a singularity, and calculate how much time must have passed between that moment (the Big Bang) and the present.

Methods for Measuring the Age of the Universe

The better method that Siegel discussed is the combination of several different kinds of data, such as the current measurement of the expansion rate of the universe, changes in the rate of expansion from type Ia supernovae, clumping of matter on large scales, and fluctuations in the cosmic microwave background. When these data are combined with a model of the big bang, cosmologists can compute an age for the universe. It is important to note that the computed age for the universe depends upon the particular version of the big bang model that is assumed: if the model changes, the age changes. The current estimated age is 13.81 billion years, with an uncertainty of 120 million years. One hundred twenty million years may sound like a lot of time, but when compared with 13.81 billion years, it is within one percent. In commenting on this precision, Siegel wrote, "We have a number of different data sets that point to this conclusion, but in reality, it's all the same method." Again, this is unusually honest because many others attempt to expand this to be several different methods.

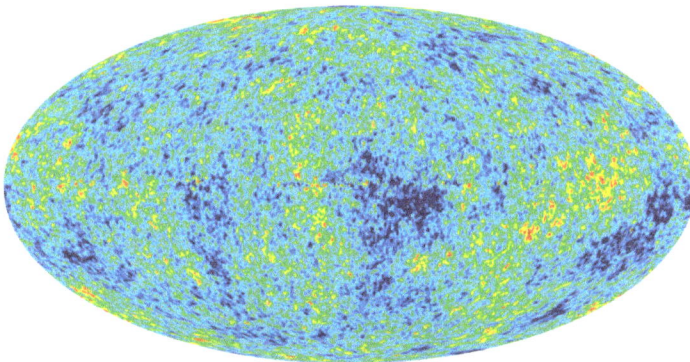

Cosmic Microwave Background.

Siegel discusses using the ages of globular star clusters as a second, less accurate method of measuring the age of the universe. A globular cluster is a centrally condensed, radially symmetric, gravitationally bound group of stars that contains between 50,000 and a half-million stars. Our galaxy, the Milky Way, contains about 200 globular clusters that orbit in the halo, a spherically symmetrical region around the galaxy. Globular clusters are thought to consist of some of the oldest stars in the galaxy, dating back to the beginning of the Milky Way, shortly after the big bang. The age estimates of globular clusters come from comparing observations of their stars (what astronomers call color-magnitude diagrams) to the predicted behavior of stars over time. Stars are lit by thermonuclear reactions deep within them. As stars age, their interior composition changes. These changes in composition alter the structure of stars, bringing about gradual changes in the gross properties of stars, which ought to show up in color-magnitude diagrams. By comparing observed color-magnitude diagrams of globular clusters with calculated changes over time from models of stellar evolution, astronomers expect the best fit between observations and theoretical models to reveal the ages of globular clusters. As with age estimates of the universe within the big bang model, the ages of globular clusters are model dependent. The current estimate of the age of globular clusters is about 13.2 billion years, though there is uncertainty of a billion years or so. Since this age is 600 million years less than the big bang age of 13.81 billion years, this is thought to be good confirmation of the universe's age.

Observational Limits

Since the universe must be at least as old as the oldest things in it, there are a number of observations which put a lower limit on the age of the universe; these include the temperature of the coolest white dwarfs, which gradually cool as they age, and the dimmest turnoff point of main sequence stars in clusters (lower-mass stars spend a greater amount of time on the main sequence, so the lowest-mass stars that have evolved off of the main sequence set a minimum age).

Cosmological Parameters

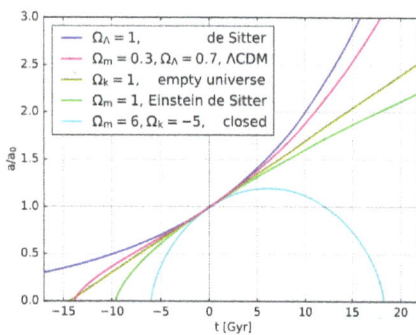

The age of the universe can be determined by measuring the Hubble constant today and extrapolating back in time with the observed value of density parameters (Ω).

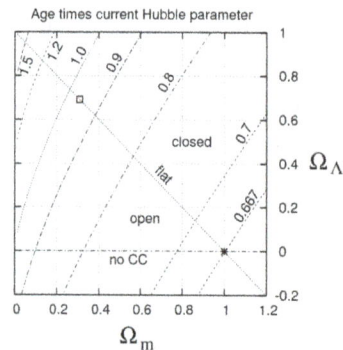

The value of the age correction factor, F, is shown as a function of two cosmological parameters: the current fractional matter density Ω_m and cosmological constant density Ω_Λ.

The problem of determining the age of the universe is closely tied to the problem of determining the values of the cosmological parameters. Today this is largely carried out in the context of the ΛCDM model, where the universe is assumed to contain normal (baryonic) matter, cold dark matter, radiation (including both photons and neutrinos), and a cosmological constant. The fractional contribution of each to the current energy density of the universe is given by the density parameters Ωm, Ωr, and $\Omega\Lambda$. The full ΛCDM model is described by a number of other parameters, but for the purpose of computing its age these three, along with the Hubble parameter H_0, are the most important.

If one has accurate measurements of these parameters, then the age of the universe can be determined by using the Friedmann equation. This equation relates the rate of change in the scale factor $a(t)$ to the matter content of the universe. Turning this relation around, we can calculate the change in time per change in scale factor and thus calculate the total age of the universe by integrating this formula. The age t_0 is then given by an expression of the form

$$t_0 = \frac{1}{H_0} F(\Omega_r, \Omega_m, \Omega_\Lambda, \ldots)$$

where H_0 is the Hubble parameter and the function F depends only on the fractional contribution to the universe's energy content that comes from various components. The first observation that one can make from this formula is that it is the Hubble parameter that controls that age of the universe, with a correction arising from the matter and energy content. So a rough estimate of the age of the universe comes from the Hubble time, the inverse of the Hubble parameter. With a value for H_0 around 68 km/s/Mpc, the Hubble time evaluates to $1/H_0 = 14.4$ billion years.

To get a more accurate number, the correction factor F must be computed. In general this must be done numerically, and the results for a range of cosmological parameter values are shown in the figure. For the Planck values $(\Omega_m, \Omega_\Lambda) = (0.3086, 0.6914)$, shown by the box in the upper left corner of the figure, this correction factor is about $F = 0.956$. For a flat universe without any cosmological constant, shown by the star in the lower right corner, $F = \frac{2}{3}$ is much smaller and thus the universe is younger for a fixed value of the Hubble parameter. To make this figure, Ω_r is held constant (roughly equivalent to holding the CMB temperature constant) and the curvature density parameter is fixed by the value of the other three.

Apart from the Planck satellite, the Wilkinson Microwave Anisotropy Probe (WMAP) was instrumental in establishing an accurate age of the universe, though other measurements must be folded in to gain an accurate number. CMB measurements are very good at constraining the matter content Ω_m and curvature parameter Ω_k. It is not as sensitive to Ω_Λ directly, partly because the cosmological constant becomes important only at low redshift. The most accurate determinations of the Hubble parameter H_0 come from Type Ia supernovae. Combining these measurements leads to the generally accepted value for the age of the universe quoted above.

The cosmological constant makes the universe "older" for fixed values of the other parameters. This is significant, since before the cosmological constant became generally accepted, the Big Bang model had difficulty explaining why globular clusters in the Milky Way appeared to be far older than the age of the universe as calculated from the Hubble parameter and a matter-only universe. Introducing the cosmological constant allows the universe to be older than these clusters, as well as explaining other features that the matter-only cosmological model could not.

WMAP

NASA's Wilkinson Microwave Anisotropy Probe (WMAP) project's nine-year data release in 2012 estimated the age of the universe to be $(13.772 \pm 0.059) \times 10^9$ years (13.772 billion years, with an uncertainty of plus or minus 59 million years).

However, this age is based on the assumption that the project's underlying model is correct; other methods of estimating the age of the universe could give different ages. Assuming an extra background of relativistic particles, for example, can enlarge the error bars of the WMAP constraint by one order of magnitude.

This measurement is made by using the location of the first acoustic peak in the microwave background power spectrum to determine the size of the decoupling surface (size of the universe at the time of recombination). The light travel time to this surface (depending on the geometry used) yields a reliable age for the universe. Assuming the validity of the models used to determine this age, the residual accuracy yields a margin of error near one percent.

Planck

In 2015, the Planck Collaboration estimated the age of the universe to be 13.813±0.038 billion years, slightly higher but within the uncertainties of the earlier number derived from the WMAP data. By combining the Planck data with external data, the best combined estimate of the age of the universe is $(13.799\pm0.021)\times10^9$ years old.

Cosmological parameters from Planck results

Parameter	Symbol	TT+lowP 68% limits	TT+lowP +lensing 68% limits	TT+lowP +lensing+ext 68% limits	TT,TE,EE+ lowP 68% limits	TT,TE,EE+ lowP + lensing 68% limits	TT,TE,EE+lowP +lensing+ext 68% limits
Age of the universe (Ga)	t_0	13.813±0.038	13.799±0.038	13.796±0.029	13.813±0.026	13.807±0.026	13.799±0.021
Hubble constant ($^{km}/_{Mpc\cdot s}$)	H_0	67.31±0.96	67.81±0.92	67.90±0.55	67.27±0.66	67.51±0.64	67.74±0.46

68% limits: Parameter 68% confidence limits for the base ΛCDM model

TT, TE, EE: Planck Cosmic microwave background (CMB) power spectra

lowP: Planck polarization data in the low-ℓ likelihood

lensing: CMB lensing reconstruction

ext: External data (BAO+JLA+H0). BAO: Baryon acoustic oscillations, JLA: Joint Light-curve Analysis, H0: Hubble constant

Assumption of strong priors

Calculating the age of the universe is accurate only if the assumptions built into the models being used to estimate it are also accurate. This is referred to as strong priors and essentially involves stripping the potential errors in other parts of the model to render the accuracy of actual observational data directly into the concluded result. Although this is not a valid procedure in all contexts (as noted in the accompanying caveat: "based on the fact we have assumed the underlying model we used is correct"), the age given is thus accurate to the specified error (since this error represents the error in the instrument used to gather the raw data input into the model).

The age of the universe based on the best fit to Planck 2015 data alone is 13.813±0.038 billion years (the estimate of 13.799±0.021 billion years uses Gaussian priors based on earlier estimates from other studies to determine the combined uncertainty). This number represents an accurate "direct" measurement of the age of the universe (other methods typically involve Hubble's law and the age of the oldest stars in globular clusters, etc.). It is possible to use different methods for determining the same parameter (in this case – the age of the universe) and arrive at different answers with no overlap in the "errors". To best avoid the problem, it is common to show two sets of uncertainties; one related to the actual measurement and the other related to the systematic errors of the model being used.

An important component to the analysis of data used to determine the age of the universe (e.g. from Planck) therefore is to use a Bayesian statistical analysis, which normalizes the results based upon the priors (i.e. the model). This quantifies any uncertainty in the accuracy of a measurement due to a particular model used.

Shape of the Universe

The shape of the Universe is a subject of investigation within physical cosmology.

Cosmologists and astronomers describe the geometry of the Universe which includes both local geometry and global geometry.

The shape of the universe can be determined by measuring the average density of matter within it, assuming that all matter is evenly distributed, rather than the distortions caused by 'dense' objects such as galaxies.

This assumption is justified by the observations that, while the universe is "weakly" inhomogeneous and anisotropic, it is on average homogeneous and isotropic.

Considerations of the geometry of the universe can be split into two parts; the local geometry relates to the observable universe, while the global geometry relates to the universe as a whole - including that which we can't measure.

The Geometry of the Cosmos

According to Einstein's theory of General Relativity, space itself can be curved by mass. As a result, the density of the universe — how much mass it has spread over its volume — determines its shape, as well as its future.

Scientists have calculated the "critical density" of the universe. The critical density is proportional to the square of the Hubble constant, which is used in measuring the expansion rate of the universe. Comparing the critical density to the actual density can help scientists to understand the cosmos.

If the actual density of the universe is less than the critical density, then there is not enough matter to stop the expansion of the universe, and it will expand forever. The resulting shape is curved like the surface of a saddle. This is known as an open universe.

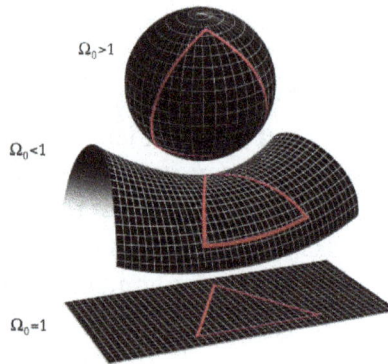

The shape of the universe depends on its density. If the density is more than the critical density, the universe is closed and curves like a sphere; if less, it will curve like a saddle. But if the actual density of the universe is equal to the critical density, as scientists think it is, then it will extend forever like a flat piece of paper.

If the actual density of the universe is greater than the critical density, then it contains enough mass to eventually stop its expansion. In this case, the universe is closed and finite, though it has no end, and has a spherical shape. Once the universe stops expanding, it will begin to contract. Galaxies will stop receding and start moving closer and closer together. Eventually, the universe will undergo the opposite of the Big Bang, often called the "Big Crunch." This is known as a closed universe.

However, if the universe contains exactly enough mass to eventually stop the expansion, the actual density of the universe will equal the critical density. The expansion rate will slow down gradually, over an infinite amount of time. In such a case, the universe is considered flat and infinite in size.

Measurements indicate that the universe is flat, suggesting that it is also infinite in size. The speed of light limits us to viewing the volume of the universe visible since the Big Bang; because the universe is approximately 13.8 billion years old, scientists can only see 13.8 billion light-years from Earth.

References

- The Compact Edition of the Oxford English Dictionary. II. Oxford: Oxford University Press. 1971. pp. 569, 909, 1900, 3821–22. ISBN 978-0198611172

- C. Sivaram (1986). "Evolution of the Universe through the Planck epoch". Astrophysics & Space Science. 125: 189–199. Bibcode:1986Ap&SS.125..189S. doi:10.1007/BF00643984

- Edward Robert Harrison (2000). Cosmology: the science of the universe. Cambridge University Press. pp. 447–. ISBN 978-0-521-66148-5. Retrieved May 1, 2011

- Piero Madau; et al. (1999). "Radiative Transfer in a Clumpy Universe. III. The Nature of Cosmological Ionizing Source". The Astrophysical Journal. 514 (2): 648–659. arXiv:astro-ph/9809058. Bibcode:1999ApJ...514..648M. doi:10.1086/306975

- Schutz, Bernard (May 31, 2009). A First Course in General Relativity (2 ed.). Cambridge University Press. p. 142 & 171. ISBN 0-521-88705-4

- Munitz MK (1959). "One Universe or Many?". Journal of the History of Ideas. 12 (2): 231–55. doi:10.2307/2707516. JSTOR 2707516

- "What is the Ultimate Fate of the Universe?". National Aeronautics and Space Administration. NASA. Retrieved August 23, 2015

- Nickolay Gnedin & Jeremiah Ostriker (1997). "Reionization of the Universe and the Early Production of Metals". Astrophysical Journal. 486 (2): 581–598. arXiv:astro-ph/9612127. Bibcode:1997ApJ...486..581G. doi:10.1086/304548

Chapter 2

An Introduction to Cosmology

The study of the origin, evolution and the eventual fate of the universe is under the domain of cosmology. This chapter discusses in detail the theories and models central to the development of cosmology, such as Big Bang theory, Lambda-CDM model, cosmological constant and recombination, among others.

Cosmology is the study of the Universe and its components, how it formed, how its has evolved and what is its future. Modern cosmology grew from ideas before recorded history. Ancient man asked questions such as "What's going on around me?" which then developed into "How does the Universe work?", the key question that cosmology asks.

To religious studies, cosmology is about a theistically created world ruled by supernatural forces. To scientists, modern cosmology is about developing the most complete and economical as possible understanding of the Universe that is consistent with observations elucidated by natural forces. We will primarily explore the latter type of cosmology in this course.

Many of the earliest recorded scientific observations were about cosmology, and pursue of understanding has continued for over 5000 years. Cosmology has exploded in the last 20 years with radically new information about the structure, origin and evolution of the Universe obtained through recent technological advances in telescopes and space observatories and basically has become a search for the understanding of not only what makes up the Universe (the objects within it) but also its overall architecture.

Modern cosmology is on the borderland between science and philosophy, close to philosophy because it asks fundamental questions about the Universe, close to science since it looks for answers in the form of empirical understanding by observation and rational explanation. Thus, theories about cosmology operate with a tension between a philosophical urge for simplicity and a wish to include all the Universe's features versus the total complexity of it all.

Neolithic Cosmology

Cosmology is as old as humankind. Once primitive socal groups developed language, it was a short step to making their first attempts to understand the world around them. Very early cosmology, from Neolithic times of 20,000 to 100,000 years ago, was extremely local. The Universe was what you immediately interacted with. Cosmological things were weather, earthquakes, sharp changes in your environment, etc. Things outside your daily experience appeared supernatural, and so we call this the time of Magic Cosmology.

The earliest physical evidence of astronomical and cosmological thinking is a lunar calendar found on a bone fragment in Sub-Saharan Africa dated at about 20,000 BC. Late megalithic structures with astronomical purpose appear in Africa and Europe around 5,000 BC (primitive versions of the famous Stonehenge complex in Britain). It is important to note that these structures and technologies were constructed by numerous different cultures which had had no contact with each other. In other words, the conclusions they reached about the cosmos were universal and the people of time were willing to commit significant resources to express these ideas.

Early people projected their own inner thoughts and feelings into an outer animistic world, a world where everything was alive. Through prayers, sacrifices and gifts to the spirits, human beings gained control of the phenomena of their world. This is an anthropomorphic (magic) worldview, of the living earth, water, wind and fire, into which men and women projected their own emotions and motives as the guiding forces, the kind of world one finds in fantasy and fairy tales.

The earliest recorded astronomical observation is the Nebra sky disk from northern Europe dating approximately 1,600 BC. This 30 cm bronze disk depicts the Sun, a lunar crescent and stars (including the Pleiades star cluster). The disk is probably a religious symbol as well as a crude astronomical instrument or calendar. In the Western hemisphere, similar understanding of basic stellar and planetary behavior was developing. For example, Native American culture around the same time were leaving rock drawings, or petroglyphs, of astronomical phenomenon. The clearest example is found below, a petroglyph which depicts the 1,006 AD supernova that resulted in the Crab Nebula.

Later in history, 5,000 to 20,000 years ago, humankind begins to organize themselves and develop what we now call culture. A greater sense of permanence in your daily existences leads to the development of myths, particularly creation myths to explain the origin of the Universe.

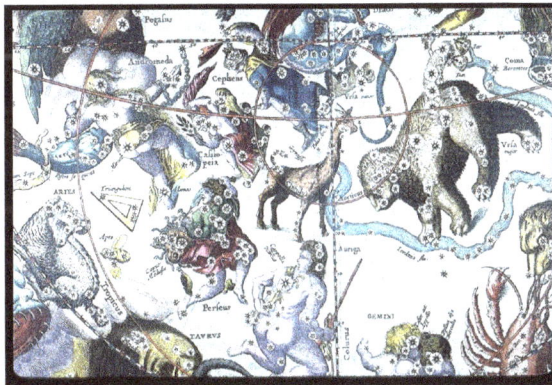

Most myths maintain supernatural themes, with gods, divine and semi-divine figures, but there was usually an internal logical consistence to the narrative. Myths are often attempts at a rational explanation of the everyday world, their goal is to teach. Even if we consider some of the stories to be ridiculous, they were, in some sense, our first scientific theories. They also closely follow a particular religion, and this time is characterized by a close marriage of science and religion. We call this the time of Mythical Cosmology.

Egyptian/Mesopotamia Cosmology:

In the region now know as the Mideast, about 5,000 years ago humankind begins to organize themselves and develop the great Bronze Age cultures. These first great civilizations (clustered about the Nile and Euphrates rivers) gave the common man a greater sense of permanence to their daily existences. This leads to the development of myths, particularly creation myths, to explain the origin of the Universe. Today, we divide the study of cosmology into cosmogony, the study of the creation of the Universe, versus cosmology, the study of the structure and evolution of the Universe (and its destiny). Many of these early creation myths were first attempts at a logical, consistent cosmogony.

Historians tend to exaggerate the capabilities of ancient Egyptians, when, in fact, they were a practical culture. The development of cosmology in ancient Egypt followed practical lines. Early man's impressions of the night sky formulated into various myths which then later became the core of Egyptian religion. Since its principal deities were heavenly bodies, a great deal of effort was made by the priesthood to calculate and predict the time and place of their god's appearances. These skills led to the division of the day and night into twelve sections each, the development of a lunar calendar and the development of a solar calendar of 12 30-day months with a special 5-day unit to bring the total to 365 days.

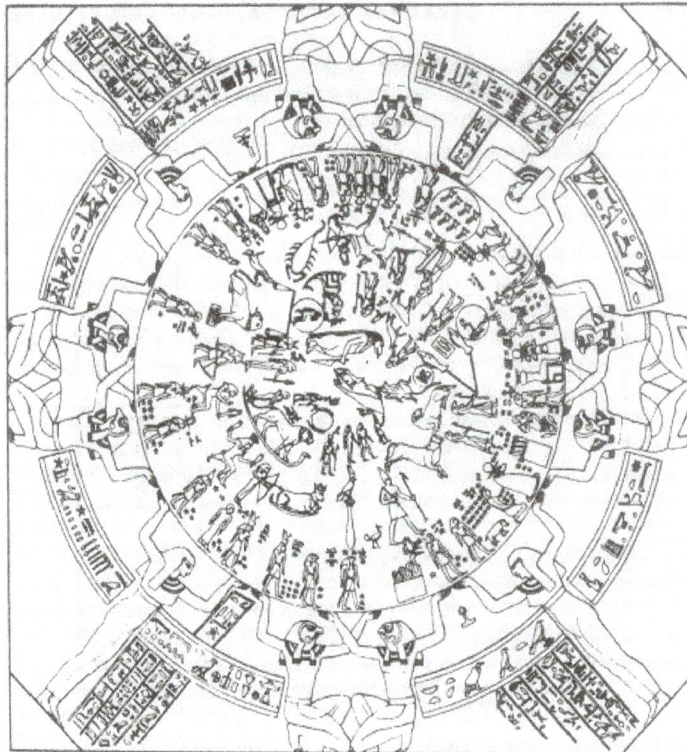

Early Egyptians impressions of the night sky formulated into various myths which then later became the core of Egyptian religion. Since its principal deities were heavenly bodies, a great deal of effort was made by the priesthood to calculate and predict the time and place of their god's appearances. Because the sun god, Ra, was the pre-eminent god, the annual solar motion along the horizon was a key astronomical observation for the Egyptians. The timing and position of the northernmost and southernmost turning points, the solstices, ultimately fixed the mythology of Egyptian cosmology. Egyptian legend declares that the sky goddess Nut gives birth to Ra once a year, catalysing both calendar development and the concept of divine royalty plus the matrilineal inheritance of the throne.

Nut is often portrayed as a naked female stretched across the sky. The Sun (Ra) is shown entering her mouth, passing through her star speckled body and emerging from her birth canal nine months later (from the spring equinox to the winter solstice). Thus, Ra becomes a self-creating god, i.e. the Universe is self-creating and eternal.

By the Old Kingdom, the astronomical/religious zeal of the pharaohs is reflected in the construction of massive pyramids at Giza. Their shape reflects the manner in which clouds and dust scatter sunlight into broad swaths forming stairways to heaven. These were stone pathways to the gods and were oriented to reach the immortal ones, i.e. the northern circumpolar stars.

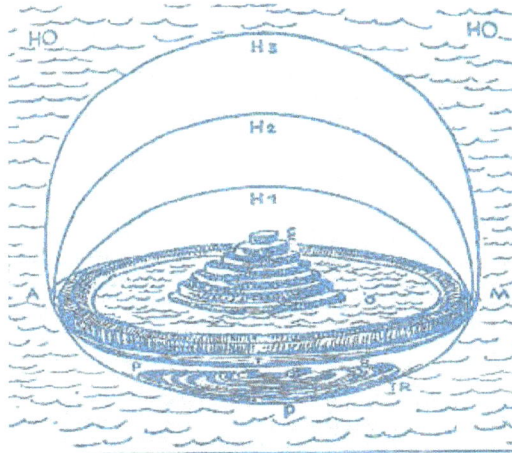

Perhaps the earliest of the combination of mythical and theological ideas is found in the Mesopotamian civilizations (divided into the Old Babylonian, Assyrian, New Babylonian and Late Babylonian periods). The Babylonian myths centered on plurality of the heavens and earth with a six-level universe consisting of three heavens and three earths: two heavens above the sky, the heaven of the stars, the earth, the underground of the Apsu, and the underworld of the dead. The Earth was created by the god Marduk as a raft floating on fresh water (Apsu) surrounded by a vastly larger body of salt water (Tiamat). The gods were divided into two pantheons, one occupying the heavens and the other in the underworld.

Babylonian astronomy is noted for their detailed, and continuous, records of astronomical phenomenon such as eclipses, positions of the planets and rise and setting of the Moon. These records date back to 800 B.C. and are the oldest scientific documents in existence. The purpose of this activity was clearly astrological with the aim of forecasting the fortunes of the country as well as of the king. In addition to records, Babylonian astronomers also developed several arithmetic tools to aid in the prediction of eclipses and planetary motion. However, while their record keeping was a novel technology for the time, and their system of stellar names and measurement system was passed onto later civilizations, the Babylonians never developed a cosmological model in which to interpret their observations. Greek astronomers will achieve this goal using the Babylonian data.

Greek Cosmology

Perhaps one of the greatest influences on modern thought are the ideas that arose from Greek philosophy between 600 BC and start of the Roman Empire. The works of scholars from this era will influence philosophers and scientists into the 21st century and many of our modern cosmological frameworks have their root in ancient Greek ideas. While many of our first cosmologies were based on myths and legends, it is the Greek philosophical tradition that introduces an intellectual approach based on evidence, reason and debate. While many of their ideas barely qualify as scientific theories, their reliance on mathematics as a tool to understand the Universe remains to this day.

The third stage, what makes up the core of modern cosmology, grew out of ancient Greek, later adopted by the Church. The underlying theme in Greek science is the use of observation and experimentation to search for simple, universal laws. We call this the time of Geometric Cosmology.

The struggle to formulation a Geometric Cosmology led to the development of the biggest philosophical achievement of humankind, the philosophy of science. Indirectly, through an examination of our myths and creation stories, we developed the ideas and techniques that later would become the core ideas to this thing we call science.

Central to Greek cosmology is the belief that the underlying order of the Universe can be expressed in mathematical form lies at the heart of science and is rarely questioned. But is mathematics a human invention or does it have an independent existence?

Idealization of physical phenomenon led Plato to hypothesize that there were two Universes, the physical world and an immaterial world of `forms', perfect aspects of everyday things such as a table, bird, and ideas/emotions, joy, action, etc. The objects and ideas in our material world are `shadows' of the forms. This solves the problem of how objects in the material world are all distinct (no two tables are exactly the same) yet they all have `tableness' in common.

Scientific Cosmology

Also called physical cosmology, it is the branch of science that deals with the scientific study of the origins and evolution of the Universe and the nature of the Universe on its very largest scales.

In its earliest form physical cosmology was basically just celestial mechanics, the study of the 'heavens'. The Greek philosophers Aristarchus of Samos, Aristotle and Ptolemy proposed different theories for how the heavens work.

In particular, the 'earth-centric' Ptolemaic system with it's perfect circles and small 'epicycles' was the accepted theory to explain the motion of the heavenly bodies. In the picture above, the Earth is at the centre with the Sun, Moon and planets all revolving around Earth.

It was the best theory until Copernicus, Kepler and Galileo proposed a 'sun-centric' system in the 16th century. Although Greek, Indian and Muslim savants formulated the sun-centric theory centuries before Copernicus, his reiteration that the Sun, rather than the Earth, is at the center of the solar system is considered among the most important landmarks in the history of modern astronomy.

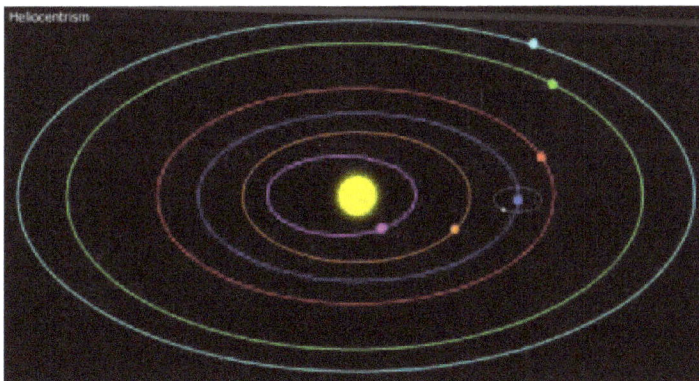

Newton's Cosmology

With Isaac Newton's 1687 publication of 'Principia Mathematica', the problem of the motion of the heavenly bodies was solved. Newton provided a physical mechanism for Kepler's laws of planetary motion. His law of universal gravitation resolved the anomalies caused by gravitational interaction between the planets in the previous systems.

Despite the words 'universal gravitation', Newton did not consider the implications of his theory on the Universe at large, although it implied that gravity works the same everywhere. The universal application of gravity was left to Albert Einstein, who formulated it more than two centuries later.

Heber D. Curtis, on the other hand, suggested that the observed spiral nebulae were star systems in their own right, or 'island universes'. In 1923-24 Edwin Hubble detected novae in the Andromeda galaxy and then showed its distance to be way beyond the Milky Way's boundaries. This settled the debate and it was accepted that the Milky Way was not all the universe there is. However, the cosmos was still thought to be static and unchanging.

Modern Cosmology

Subsequent modeling of the universe explored the possibility that the cosmological constant introduced by Einstein may result in an expanding universe, depending on its value. In 1929, Edwin Hubble's discovered the red shift of the light of distant galaxies, indicating that they move away from the Milky Way. Hence, the universe must be expanding.

Einstein's Cosmology

Scientific cosmology really began in 1917, when Albert Einstein's published the final modification to his theory of gravity in the paper 'Cosmological Considerations of the General Theory of Relativity'. This paper prompted early cosmologists such as Willem de Sitter, Karl Schwarzschild and Arthur Eddington to explore the astronomical consequences of the theory of relativity.

Mount Wilson astronomer Harlow Shapley championed the model of a cosmos made up of the Milky Way star system only – a static unchanging universe. Einstein also believed this and 'fudged' his equations of general relativity to represent this static state. He introduced a 'cosmological constant' that prevented the universe from contracting or expanding, which was what his original equations told him.

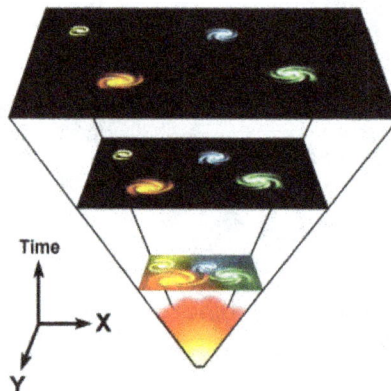

Einstein promptly repudiated his earlier 'fudge factor' and declared that the introduction of the cosmological constant was the "biggest blunder of his scientific career". His original 1917 equations support an expanding universe without the cosmological constant anyway.

In 1931 Georges Lemaitre, a Belgian priest, postulated that an expanding universe meant a creation event at some time in the past. He postulated that the Universe was created from a small 'cosmic egg' and set to expand until today. This theory was later labeled the 'big bang'.

There were many rival theories to the big bang. The 1964 discovery of the cosmic microwave background radiation by Arno Penzias and Robert Woodrow Wilson was a first step in ruling out many alternative physical cosmologies.

The Golden Age of Cosmology

Recent observations made by the COBE and WMAP satellites observing and accurately measuring this background radiation have effectively, transformed cosmology from a highly speculative science into a predictive science. This has led many to refer to modern times as the "golden age of cosmology".

Observational Cosmology

Understanding the nature and origin of large-scale structure in the Universe is one of most compelling issues in observational cosmology. The currently most conventional scenario is given by the cold dark matter (CDM) dominated model, where gravitational instability mainly driven by spatial inhomogeneities of CDM distribution amplifies the seed density perturbations to form the present-day hierarchical structures. Therefore revealing distribution and amount of CDM is crucial to understanding the formation of large-scale structure. In addition the presence of dark energy drives the accelerating cosmic expansion, and therefore affects the growth of structure formation. The dark matter distribution and the nature of dark energy can be explored from massive galaxy surveys.

We have been actively working both on the measurements using currently available facilities and on the planning of future instruments. The powerful investigative tools are the gravitational lensing effect and the baryon acoustic oscillation. Also, the cosmic microwave background (CMB) polarization is a unique tool to investigate the universe in other aspects.

Gravitational Lensing Effect

The path of light ray emitted by a distant galaxy is bent by gravitational force of intervening large-scale structure during the propagation, causing the image to be distorted—the so-called weak lensing shear. Conversely, measuring the coherent shear signals between galaxy images allows us to reconstruct the distribution of invisible dark matter. Moreover, since the weak lensing shear deals with the light propagation on cosmological distance scales, the lensing strengths depend on the cosmic expansion history that is sensitive to the nature of dark energy. Thus weak lensing based observables offer a powerful way for studying the nature of invisible components, dark matter and dark energy. We are carrying out observational and theoretical studies of weak lensing phenomena using our own Subaru data sets as well as simulations of large-scale structure.

Baryon Acoustic Oscillation

To measure properties of dark energy, one needs to measure the expansion history of the universe precisely. Because light travels at a finite speed, one can measure the expansion rate of the past by looking far. Comparing the expansion rate at varying distances would reveal the expansion history. The expansion itself is relatively easy to measure. The light emitted by a distant galaxy is stretched by the expansion of space and becomes redder, which can be measured by any decent spectrograph.

To measure the expansion history, however, we also need to know how far back in time the light was emitted from the galaxy, or equivalently, how far away it is. Measuring precise distances in cosmological scales is very challenging. Clustering of baryonic matter at a certain characteristic scale that is imprinted by baryon acoustic oscillation (BAO), or propagation of acoustic waves, in the early universe serves as a "standard ruler" for cosmological observations. This technique requires to study millions of galaxies in a wide field of view, and map the spatial distribution of luminous galaxies to detect the characteristic scale.

Hyper Suprime-Cam (HSC)

TThe HSC replaced the prime focus camera of Subaru Telescope (8.2 meter optical-infrared telescope at the summit of 4,200m-Mauna Kea, Hawaii) with a new camera that has wider field-of-view than the previous one by a factor of 10 (nine times the area of a full moon). Fully utilizing the unique capabilities of HSC, its survey speed and excellent image quality, we have conducted a massive galaxy survey from 2014 that covers an area of about thousand square degrees (equivalent to more than 5000 times the area of a full moon) and reaches to the depth to probe the Universe up to redshifts of a few. These provide us ideal data sets for exploring the nature of dark matter and dark energy via measurements of cosmological observables available from the data, weak lensing and galaxy clustering statistics. Kavli IPMU members, are actively involved in this HSC project, and working on the designing and planning of HSC galaxy survey and development of data analysis pipeline.

Sloan Digital Sky Survey IV (SDSS-IV)

The Kavli IPMU is a full member of the SDSS-IV collaboration. SDSS-IV is currently conducting the expanded Baryon Oscillation Spectroscopic Survey (eBOSS) to map the distribution of luminous galaxies and quasars throughout the universe. By focusing on redshift regimes currently unexplored by other surveys, the full eBOSS data set will probe structure over 80% of cosmic history and create the largest volume map of the universe to date. These data are being used to understand the nature of dark energy and structure formation over time. Kavli IPMU researchers also continue to make use of the legacy imaging and spectroscopic data from earlier generations of the SDSS collaboration.

Prime Focus Spectrograph (PFS)

Since Jan 2011 when PFS was approved by the community at the Subaru users meeting and the project formally started subsequently, the instrument development and large survey observation planning have been promoted by an international collaboration at the initiative of Kavli IPMU.

This instrument enables to observe max. ~2400 astronomical objects and deliver their spectra over the wavelengths from 380nm to 1260nm in one exposure. PFS on the Subaru telescope is truly a unique multipurpose spectrometer that maximally exploits the Subaru's large light-gathering power of the 8.2m-diameter primary mirror and the ultra wide field at its prime focus.

At the Subaru's prime focus, Hyper Suprime-Cam (HSC) has been operating and delivering superb images over a large area of the sky. PFS and HSC are the dual wheels of the SuMIRe (Subaru Measurement of Images and Redshifts) project to conduct a large census of the Universe. Extremely large statistics of stars and galaxies with HSC images and PFS spectra enable to provide physical constraints on dark matter, dark energy, and neutrino mass from various precision measurements such as Baryon Acoustic Oscillation, and to address the physical processes that originate the large variety of galaxy populations.

Cosmic Microwave Background (CMB) Experiments

The cosmic inflation is one of the biggest questions to be addressed in cosmology. Thankfully the event at 10^{-38} sec after the beginning of the universe can be experimentally testable by using the cosmic microwave background (CMB) polarization, called primordial B-mode. The past CMB experiments have been providing data to study the physics of early universe. The recent data from the European Space Agency (ESA) satellite Planck has solidified our understanding of Lambda-CDM cosmology. With a rapid technological development for higher sensitivity observations, the community has begun moving even to a deeper sensitivity for the direct evidence of the inflationary paradigm. In addition to the inflationary universe, there is rich physics to learn in the late universe. The CMB photon interacts with the gravitational potential from the large-scale structure, and thus anything that influences to the large-scale structure, such as the total mass of the neutrino, can be probed by the CMB polarization at the fine angular resolution.

The Kavli IPMU is actively involved to the observational efforts that are in progress from the ground and in preparation for a future satellite mission. POLARBEAR has been observing from Atacama, Chile. It delivered the first detection of the CMB B-mode power spectrum originated from the weak gravitational lensing. POLARBEAR2, a part of the Simons Array, is soon to be deployed and starts observing. Simons Observatory is the next generation ground base telescope in Chile. It probes the angular scale from the recombination dump to the weak gravitational lensing in the CMB B-mode power spectrum with high accuracy. The next generation satellite proposal, LiteBIRD, is specifically designed to aim the primordial B-mode signal with the sensitivity of the tensor-to-scalar ratio less than 0.001.

Currently, the CMB group at Kavli IPMU carries out the B-mode polarization experiments in theory and experiment. The 4 faculties and postdocs are actively leading on the projects.

Big Bang Theory

The Big Bang theory is an effort to explain what happened at the very beginning of our universe. Discoveries in astronomy and physics have shown beyond a reasonable doubt that our universe did in fact have a beginning. Prior to that moment there was nothing; during and after that moment there was something: our universe. The big bang theory is an effort to explain what happened during and after that moment.

According to the standard theory, our universe sprang into existence as "singularity" around 13.7 billion years ago. What is a "singularity" and where does it come from? Well, to be honest, we don't know for sure. Singularities are zones which defy our current understanding of physics. They are thought to exist at the core of "black holes." Black holes are areas of intense gravitational pressure. The pressure is thought to be so intense that finite matter is actually squished into infinite density (a mathematical concept which truly boggles the mind). These zones of infinite density are called "singularities." Our universe is thought to have begun as an infinitesimally small, infinitely hot, infinitely dense, something - a singularity. Where did it come from? We don't know. Why did it appear? We don't know.

After its initial appearance, it apparently inflated (the "Big Bang"), expanded and cooled, going from very, very small and very, very hot, to the size and temperature of our current universe. It continues to expand and cool to this day and we are inside of it: incredible creatures living on a unique planet, circling a beautiful star clustered together with several hundred billion other stars in a galaxy soaring through the cosmos, all of which is inside of an expanding universe that began as an infinitesimal singularity which appeared out of nowhere for reasons unknown. This is the Big Bang theory.

Common Misconceptions

There are many misconceptions surrounding the Big Bang theory. For example, we tend to imagine a giant explosion. Experts however say that there was no explosion; there was (and continues to be) an expansion. Rather than imagining a balloon popping and releasing its contents, imagine a balloon expanding: an infinitesimally small balloon expanding to the size of our current universe.

Another misconception is that we tend to image the singularity as a little fireball appearing somewhere in space. According to the many experts however, space didn't exist prior to the Big Bang. Back in the late '60s and early '70s, when men first walked upon the moon, "three British astrophysicists, Steven Hawking, George Ellis, and Roger Penrose turned their attention to the Theory of Relativity and its implications regarding our notions of time. In 1968 and 1970, they published papers in which they extended Einstein's Theory of General Relativity to include measurements of time and space. According to their calculations, time and space had a finite beginning that corresponded to the origin of matter and energy." The singularity didn't appear *in* space; rather, space began inside of the singularity. Prior to the singularity, *nothing* existed, not space, time, matter, or energy - nothing. So where and in what did the singularity appear if not in space? We don't know. We don't know where it came from, why it's here, or even where it is. All we really know is that we are inside of it and at one time it didn't exist and neither did we.

Evidence for the Theory

What are the major evidences which support the Big Bang theory?

- First of all, we are reasonably certain that the universe had a beginning.

- Second, galaxies appear to be moving away from us at speeds proportional to their distance. This is called "Hubble's Law," named after Edwin Hubble (1889-1953) who discovered this phenomenon in 1929. This observation supports the expansion of the universe and suggests that the universe was once compacted.

- Third, if the universe was initially very, very hot as the Big Bang suggests, we should be able to find some remnant of this heat. In 1965, Radioastronomers Arno Penzias and Robert Wilson discovered a 2.725 degree Kelvin (-454.765 degree Fahrenheit, -270.425 degree Celsius) Cosmic Microwave Background radiation (CMB) which pervades the observable universe. This is thought to be the remnant which scientists were looking for. Penzias and Wilson shared in the 1978 Nobel Prize for Physics for their discovery.

- Finally, the abundance of the "light elements" Hydrogen and Helium found in the observable universe are thought to support the Big Bang model of origins.

The Only Plausible Theory

Is the standard Big Bang theory the only model consistent with these evidences? No, it's just the most popular one. Internationally renown Astrophysicist George F. R. Ellis explains: "People need to be aware that there is a range of models that could explain the observations. For instance, I can construct you a spherically symmetrical universe with Earth at its center, and you cannot disprove it based on observations. You can only exclude it on philosophical grounds. In my view there is absolutely nothing wrong in that. What I want to bring into the open is the fact that we are using philosophical criteria in choosing our models. A lot of cosmology tries to hide that."

In 2003, Physicist Robert Gentry proposed an attractive alternative to the standard theory, an alternative which also accounts for the evidences listed above. Dr. Gentry claims that the standard Big Bang model is founded upon a faulty paradigm (the Friedmann-lemaitre expanding-space-time paradigm) which he claims is inconsistent with the empirical data. He chooses instead to base his model on Einstein's static-spacetime paradigm which he claims is the "genuine cosmic Rosetta." Gentry has published several papers outlining what he considers to be serious flaws in the standard Big Bang model. Other high-profile dissenters include Nobel laureate Dr. Hannes Alfvĭn, Professor Geoffrey Burbidge, Dr. Halton Arp, and the renowned British astronomer Sir Fred Hoyle, who is accredited with first coining the term "the Big Bang" during a BBC radio broadcast in 1950.

Timeline

Singularity

Extrapolation of the expansion of the universe backwards in time using general relativity yields an infinite density and temperature at a finite time in the past. This singularity indicates that general relativity is not an adequate description of the laws of physics in this regime. Models based on general relativity alone can not extrapolate toward the singularity beyond the end of the Planck epoch.

This primordial singularity is itself sometimes called "the Big Bang", but the term can also refer to a more generic early hot, dense phase of the universe. In either case, "the Big Bang" as an event is also colloquially referred to as the "birth" of our universe since it represents the point in history where the universe can be verified to have entered into a regime where the laws of physics as we understand them (specifically general relativity and the standard model of particle physics) work. Based on measurements of the expansion using Type Ia supernovae and measurements of temperature fluctuations in the cosmic microwave background, the time that has passed since that event — otherwise known as the "age of the universe" — is 13.799 ± 0.021 billion years. The agreement of independent measurements of this age supports the ΛCDM model that describes in detail the characteristics of the universe.

Despite being extremely dense at this time—far denser than is usually required to form a black hole—the universe did not re-collapse into a black hole. This may be explained by considering that commonly-used calculations and limits for gravitational collapse are usually based upon objects of relatively constant size, such as stars, and do not apply to rapidly expanding space such as the Big Bang.

Inflation and Baryogenesis

The earliest phases of the Big Bang are subject to much speculation. In the most common models the universe was filled homogeneously and isotropically with a very high energy density and huge temperatures and pressures and was very rapidly expanding and cooling. Approximately 10^{-37} seconds into the expansion, a phase transition caused a cosmic inflation, during which the universe grew exponentially during which time density fluctuations that occurred because of the uncertainty principle were amplified into the seeds that would later form the large-scale structure of the universe. After inflation stopped, reheating occurred until the universe obtained the temperatures required for the production of a quark–gluon plasma as well as all other elementary particles. Temperatures were so high that the random motions of particles were at relativistic speeds, and particle–antiparticle pairs of all kinds were being continuously created and destroyed in collisions. At some point, an unknown reaction called baryogenesis violated the conservation of baryon number, leading to a very small excess of quarks and leptons over antiquarks and antileptons—of the order of one part in 30 million. This resulted in the predominance of matter over antimatter in the present universe.

Cooling

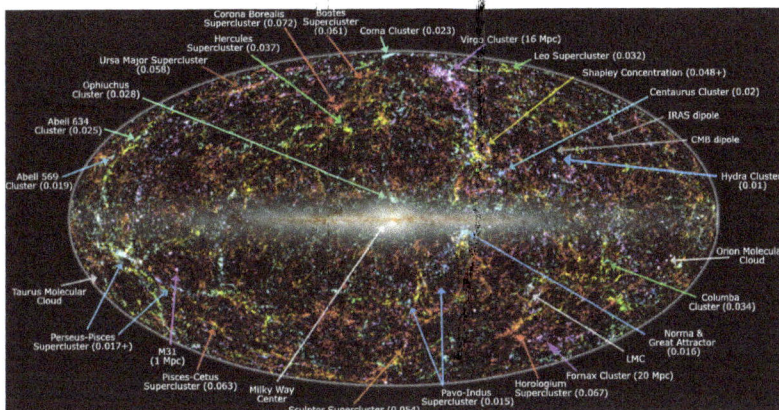

Panoramic view of the entire near-infrared sky reveals the distribution of galaxies beyond the Milky Way. Galaxies are color-coded by redshift.

The universe continued to decrease in density and fall in temperature, hence the typical energy of each particle was decreasing. Symmetry breaking phase transitions put the fundamental forces of physics and the parameters of elementary particles into their present form. After about 10^{-11} seconds, the picture becomes less speculative, since particle energies drop to values that can be attained in particle accelerators. At about 10^{-6} seconds, quarks and gluons combined to form baryons such as protons and neutrons. The small excess of quarks over antiquarks led to a small excess of baryons over antibaryons. The temperature was now no longer high enough to create new proton–antiproton pairs (similarly for neutrons–antineutrons), so a mass annihilation immediately followed, leaving just one in 10^{10} of the original protons and neutrons, and none of their antiparticles. A similar process happened at about 1 second for electrons and positrons. After these annihilations, the remaining protons, neutrons and electrons were no longer moving relativistically and the energy density of the universe was dominated by photons (with a minor contribution from neutrinos).

A few minutes into the expansion, when the temperature was about a billion (one thousand million) kelvin and the density was about that of air, neutrons combined with protons to form the universe's deuterium and helium nuclei in a process called Big Bang nucleosynthesis. Most protons remained uncombined as hydrogen nuclei.

As the universe cooled, the rest mass energy density of matter came to gravitationally dominate that of the photon radiation. After about 379,000 years, the electrons and nuclei combined into atoms (mostly hydrogen); hence the radiation decoupled from matter and continued through space largely unimpeded. This relic radiation is known as the cosmic microwave background radiation. The chemistry of life may have begun shortly after the Big Bang, 13.8 billion years ago, during a habitable epoch when the universe was only 10–17 million years old.

Structure Formation

Abell 2744 galaxy cluster – Hubble Frontier Fields view.

Over a long period of time, the slightly denser regions of the nearly uniformly distributed matter gravitationally attracted nearby matter and thus grew even denser, forming gas clouds, stars, galaxies, and the other astronomical structures observable today. The details of this process depend on the amount and type of matter in the universe. The four possible types of matter are known as cold dark matter, warm dark matter, hot dark matter, and baryonic matter. The best measurements available, from Wilkinson Microwave Anisotropy Probe (WMAP), show that the data is well-fit by a Lambda-CDM model in which dark matter is assumed to be cold (warm dark matter is ruled out by early reionization), and is estimated to make up about 23% of the matter/energy of the universe, while baryonic matter makes up about 4.6%. In an "extended model" which includes hot dark matter in the form of neutrinos, then if the "physical baryon density" $\Omega_b h^2$ is estimated at about 0.023 (this is different from the 'baryon density' Ω_b expressed as a fraction of the total matter/energy density, which as noted above is about 0.046), and the corresponding cold dark matter density $\Omega_c h^2$ is about 0.11, the corresponding neutrino density $\Omega_v h^2$ is estimated to be less than 0.0062.

Cosmic Acceleration

Independent lines of evidence from Type Ia supernovae and the CMB imply that the universe today is dominated by a mysterious form of energy known as dark energy, which apparently permeates all of space. The observations suggest 73% of the total energy density of today's universe is in this form. When the universe was very young, it was likely infused with dark energy, but with less space and everything closer together, gravity predominated, and it was slowly braking the expansion. But eventually, after numerous billion years of expansion, the growing abundance of dark energy caused the expansion of the universe to slowly begin to accelerate.

Dark energy in its simplest formulation takes the form of the cosmological constant term in Einstein's field equations of general relativity, but its composition and mechanism are unknown and, more generally, the details of its equation of state and relationship with the Standard Model of particle physics continue to be investigated both through observation and theoretically.

All of this cosmic evolution after the inflationary epoch can be rigorously described and modeled by the ΛCDM model of cosmology, which uses the independent frameworks of quantum mechanics and Einstein's General Relativity. There is no well-supported model describing the action prior to 10^{-15} seconds or so. Apparently a new unified theory of quantum gravitation is needed to break this barrier. Understanding this earliest of eras in the history of the universe is currently one of the greatest unsolved problems in physics.

Features of the Model

The Big Bang theory depends on two major assumptions: the universality of physical laws and the cosmological principle. The cosmological principle states that on large scales the universe is homogeneous and isotropic.

These ideas were initially taken as postulates, but today there are efforts to test each of them. For example, the first assumption has been tested by observations showing that largest possible deviation of the fine structure constant over much of the age of the universe is of order 10^{-5}. Also, general relativity has passed stringent tests on the scale of the Solar System and binary stars.

If the large-scale universe appears isotropic as viewed from Earth, the cosmological principle can be derived from the simpler Copernican principle, which states that there is no preferred (or special) observer or vantage point. To this end, the cosmological principle has been confirmed to a level of 10^{-5} via observations of the CMB. The universe has been measured to be homogeneous on the largest scales at the 10% level.

Expansion of Space

General relativity describes spacetime by a metric, which determines the distances that separate nearby points. The points, which can be galaxies, stars, or other objects, are themselves specified using a coordinate chart or "grid" that is laid down over all spacetime. The cosmological principle implies that the metric should be homogeneous and isotropic on large scales, which uniquely singles out the Friedmann–Lemaître–Robertson–Walker metric (FLRW metric). This metric contains a scale factor, which describes how the size of the universe changes with time. This enables a convenient choice of a coordinate system to be made, called comoving coordinates. In this coordinate system, the grid expands along with the universe, and objects that are moving only because of the expansion of the universe, remain at fixed points on the grid. While their *coordinate* distance (comoving distance) remains constant, the *physical* distance between two such co-moving points expands proportionally with the scale factor of the universe.

The Big Bang is not an explosion of matter moving outward to fill an empty universe. Instead, space itself expands with time everywhere and increases the physical distance between two co-moving points. In other words, the Big Bang is not an explosion *in space*, but rather an expansion *of space*. Because the FLRW metric assumes a uniform distribution of mass and energy, it applies to our universe only on large scales—local concentrations of matter such as our galaxy are gravitationally bound and as such do not experience the large-scale expansion of space.

Horizons

An important feature of the Big Bang spacetime is the presence of particle horizons. Since the universe has a finite age, and light travels at a finite speed, there may be events in the past whose light has not had time to reach us. This places a limit or a *past horizon* on the most distant objects that can be observed. Conversely, because space is expanding, and more distant objects are receding ever more quickly, light emitted by us today may never "catch up" to very distant objects. This defines a *future horizon*, which limits the events in the future that we will be able to influence. The presence of either type of horizon depends on the details of the FLRW model that describes our universe.

Our understanding of the universe back to very early times suggests that there is a past horizon, though in practice our view is also limited by the opacity of the universe at early times. So our view cannot extend further backward in time, though the horizon recedes in space. If the expansion of the universe continues to accelerate, there is a future horizon as well.

Observational Evidence

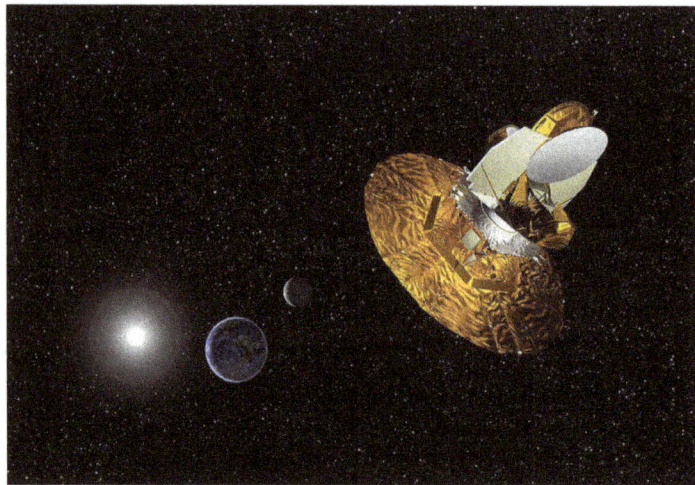

Artist's depiction of the WMAP satellite gathering data to help scientists understand the Big Bang

The earliest and most direct observational evidence of the validity of the theory are the expansion of the universe according to Hubble's law (as indicated by the redshifts of galaxies), discovery and measurement of the cosmic microwave background and the relative abundances of light elements produced by Big Bang nucleosynthesis. More recent evidence includes observations of galaxy formation and evolution, and the distribution of large-scale cosmic structures, These are sometimes called the "four pillars" of the Big Bang theory.

Precise modern models of the Big Bang appeal to various exotic physical phenomena that have not been observed in terrestrial laboratory experiments or incorporated into the Standard Model of particle physics. Of these features, dark matter is currently subjected to the most active laboratory investigations. Remaining issues include the cuspy halo problem and the dwarf galaxy problem of cold dark matter. Dark energy is also an area of intense interest for scientists, but it is not clear whether direct detection of dark energy will be possible. Inflation and baryogenesis remain more speculative features of current Big Bang models. Viable, quantitative explanations for such phenomena are still being sought. These are currently un-solved problems in physics.

Hubble's Law and the Expansion of Space

Observations of distant galaxies and quasars show that these objects are redshifted—the light emitted from them has been shifted to longer wavelengths. This can be seen by taking a frequency spectrum of an object and matching the spectroscopic pattern of emission lines or absorption lines corresponding to atoms of the chemical elements interacting with the light. These redshifts are uniformly isotropic, distributed evenly among the observed objects in all directions. If the redshift is interpreted as a Doppler shift, the recessional velocity of the object can be calculated. For some galaxies, it is possible to estimate distances via the cosmic distance ladder. When the recessional velocities are plotted against these distances, a linear relationship known as Hubble's law is observed: $v = H_0D$ where

- v is the recessional velocity of the galaxy or other distant object,

- D is the comoving distance to the object, and

- H_0 is Hubble's constant, measured to be 70.4+1.3−1.4 km/s/Mpc by the WMAP probe.

Hubble's law has two possible explanations. Either we are at the center of an explosion of galaxies—which is untenable given the Copernican principle—or the universe is uniformly expanding everywhere. This universal expansion was predicted from general relativity by Alexander Fried-mann in 1922 and Georges Lemaître in 1927, well before Hubble made his 1929 analysis and observations, and it remains the cornerstone of the Big Bang theory as developed by Friedmann, Lemaître, Robertson, and Walker.

The theory requires the relation $v = HD$ to hold at all times, where D is the comoving distance, v is the recessional velocity, and v, H, and D vary as the universe expands (hence we write H_0 to denote the present-day Hubble "constant"). For distances much smaller than the size of the observable universe, the Hubble redshift can be thought of as the Doppler shift corresponding to the recession velocity v. However, the redshift is not a true Doppler shift, but rather the result of the expansion of the universe between the time the light was emitted and the time that it was detected.

That space is undergoing metric expansion is shown by direct observational evidence of the Cos-mological principle and the Copernican principle, which together with Hubble's law have no oth-er explanation. Astronomical redshifts are extremely isotropic and homogeneous, supporting the Cosmological principle that the universe looks the same in all directions, along with much other evidence. If the redshifts were the result of an explosion from a center distant from us, they would not be so similar in different directions.

Measurements of the effects of the cosmic microwave background radiation on the dynamics of distant astrophysical systems in 2000 proved the Copernican principle, that, on a cosmological scale, the Earth is not in a central position. Radiation from the Big Bang was demonstrably warmer at earlier times throughout the universe. Uniform cooling of the CMB over billions of years is ex-plainable only if the universe is experiencing a metric expansion, and excludes the possibility that we are near the unique center of an explosion.

Cosmic Microwave Background Radiation

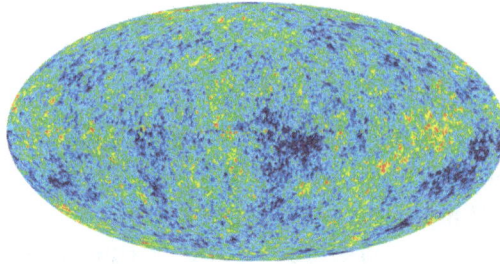

9 year WMAP image of the cosmic microwave background radiation.
The radiation is isotropic to roughly one part in 100,000.

In 1964 Arno Penzias and Robert Wilson serendipitously discovered the cosmic background radiation, an omnidirectional signal in the microwave band. Their discovery provided substantial confirmation of the big-bang predictions by Alpher, Herman and Gamow around 1950. Through the 1970s the radiation was found to be approximately consistent with a black body spectrum in all directions; this spectrum has been redshifted by the expansion of the universe, and today corresponds to approximately 2.725 K. This tipped the balance of evidence in favor of the Big Bang model, and Penzias and Wilson were awarded a Nobel Prize in 1978.

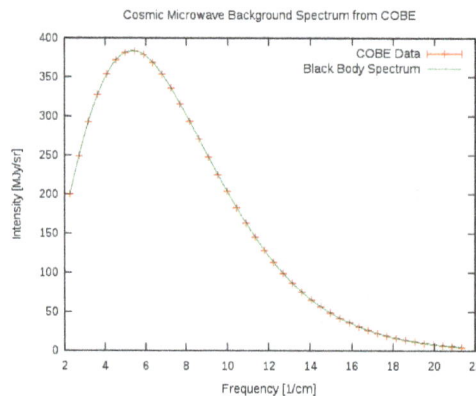

The cosmic microwave background spectrum measured by the FIRAS instrument on the
COBE satellite is the most-precisely measured black body spectrum in nature.
The data points and error bars on this graph are obscured by the theoretical curve.

The *surface of last scattering* corresponding to emission of the CMB occurs shortly after *recombination*, the epoch when neutral hydrogen becomes stable. Prior to this, the universe comprised a hot dense photon-baryon plasma sea where photons were quickly scattered from free charged particles. Peaking at around 372±14 kyr, the mean free path for a photon becomes long enough to reach the present day and the universe becomes transparent.

In 1989, NASA launched the Cosmic Background Explorer satellite (COBE), which made two major advances: in 1990, high-precision spectrum measurements showed that the CMB frequency spectrum is an almost perfect blackbody with no deviations at a level of 1 part in 10^4, and measured a residual temperature of 2.726 K (more recent measurements have revised this figure down slightly to 2.7255 K); then in 1992, further COBE measurements discovered tiny fluctuations (anisotropies) in the CMB temperature across the sky, at a level of about one part in 10^5. John C. Mather and George Smoot were awarded the 2006 Nobel Prize in Physics for their leadership in these results.

During the following decade, CMB anisotropies were further investigated by a large number of ground-based and balloon experiments. In 2000–2001 several experiments, most notably BOO-MERanG, found the shape of the universe to be spatially almost flat by measuring the typical angular size (the size on the sky) of the anisotropies.

In early 2003, the first results of the Wilkinson Microwave Anisotropy Probe (WMAP) were released, yielding what were at the time the most accurate values for some of the cosmological parameters. The results disproved several specific cosmic inflation models, but are consistent with the inflation theory in general. The Planck space probe was launched in May 2009. Other ground and balloon based cosmic microwave background experiments are ongoing.

Abundance of Primordial Elements

Using the Big Bang model it is possible to calculate the concentration of helium-4, helium-3, deuterium, and lithium-7 in the universe as ratios to the amount of ordinary hydrogen. The relative abundances depend on a single parameter, the ratio of photons to baryons. This value can be calculated independently from the detailed structure of CMB fluctuations. The ratios predicted (by mass, not by number) are about 0.25 for ^{4}He/H, about 10^{-3} for ^{2}H/H, about $10^{-4}$for ^{3}He/Hand about 10^{-9} for ^{7}Li/H.

The measured abundances all agree at least roughly with those predicted from a single value of the baryon-to-photon ratio. The agreement is excellent for deuterium, close but formally discrepant for ^{4}He, and off by a factor of two for ^{7}Li; in the latter two cases there are substantial systematic uncertainties. Nonetheless, the general consistency with abundances predicted by Big Bang nucleosynthesis is strong evidence for the Big Bang, as the theory is the only known explanation for the relative abundances of light elements, and it is virtually impossible to "tune" the Big Bang to produce much more or less than 20–30% helium. Indeed, there is no obvious reason outside of the Big Bang that, for example, the young universe (i.e., before star formation, as determined by studying matter supposedly free of stellar nucleosynthesis products) should have more helium than deuterium or more deuterium than ^{3}He, and in constant ratios, too.

Galactic Evolution and Distribution

Detailed observations of the morphology and distribution of galaxies and quasars are in agreement with the current state of the Big Bang theory. A combination of observations and theory suggest that the first quasars and galaxies formed about a billion years after the Big Bang, and since then, larger structures have been forming, such as galaxy clusters and superclusters.

Populations of stars have been aging and evolving, so that distant galaxies (which are observed as they were in the early universe) appear very different from nearby galaxies (observed in a more recent state). Moreover, galaxies that formed relatively recently, appear markedly different from galaxies formed at similar distances but shortly after the Big Bang. These observations are strong arguments against the steady-state model. Observations of star formation, galaxy and quasar distributions and larger structures, agree well with Big Bang simulations of the formation of structure in the universe, and are helping to complete details of the theory.

Primordial Gas Clouds

Focal plane of BICEP2 telescope under a microscope - used to search for polarization in the CMB.

In 2011, astronomers found what they believe to be pristine clouds of primordial gas by analyzing absorption lines in the spectra of distant quasars. Before this discovery, all other astronomical objects have been observed to contain heavy elements that are formed in stars. These two clouds of gas contain no elements heavier than hydrogen and deuterium. Since the clouds of gas have no heavy elements, they likely formed in the first few minutes after the Big Bang, during Big Bang nucleosynthesis.

Other Lines of Evidence

The age of the universe as estimated from the Hubble expansion and the CMB is now in good agreement with other estimates using the ages of the oldest stars, both as measured by applying the theory of stellar evolution to globular clusters and through radiometric dating of individual Population II stars.

The prediction that the CMB temperature was higher in the past has been experimentally supported by observations of very low temperature absorption lines in gas clouds at high redshift. This prediction also implies that the amplitude of the Sunyaev–Zel'dovich effect in clusters of galaxies does not depend directly on redshift. Observations have found this to be roughly true, but this effect depends on cluster properties that do change with cosmic time, making precise measurements difficult.

Future Observations

Future gravitational waves observatories might be able to detect primordial gravitational waves, relics of the early universe, up to less than a second after the Big Bang.

Problems and Related Issues in Physics

As with any theory, a number of mysteries and problems have arisen as a result of the development of the Big Bang theory. Some of these mysteries and problems have been resolved while others are still outstanding. Proposed solutions to some of the problems in the Big Bang model have revealed new mysteries of their own. For example, the horizon problem, the magnetic monopole problem, and the flatness problem are most commonly resolved with inflationary theory, but the details of the inflationary universe are still left unresolved and many, including some founders of the theory, say it has been disproven. What follows are a list of the mysterious aspects of the Big Bang theory still under intense investigation by cosmologists and astrophysicists.

Baryon Asymmetry

It is not yet understood why the universe has more matter than antimatter. It is generally assumed that when the universe was young and very hot it was in statistical equilibrium and contained equal numbers of baryons and antibaryons. However, observations suggest that the universe, including its most distant parts, is made almost entirely of matter. A process called baryogenesis was hypothesized to account for the asymmetry. For baryogenesis to occur, the Sakharov conditions must be satisfied. These require that baryon number is not conserved, that C-symmetry and CP-symmetry are violated and that the universe depart from thermodynamic equilibrium. All these conditions occur in the Standard Model, but the effects are not strong enough to explain the present baryon asymmetry.

Dark Energy

Measurements of the redshift–magnitude relation for type Ia supernovae indicate that the expansion of the universe has been accelerating since the universe was about half its present age. To explain this acceleration, general relativity requires that much of the energy in the universe consists of a component with large negative pressure, dubbed "dark energy".

Dark energy, though speculative, solves numerous problems. Measurements of the cosmic microwave background indicate that the universe is very nearly spatially flat, and therefore according to general relativity the universe must have almost exactly the critical density of mass/energy. But the mass density of the universe can be measured from its gravitational clustering, and is found to have only about 30% of the critical density. Since theory suggests that dark energy does not cluster in the usual way it is the best explanation for the "missing" energy density. Dark energy also helps to explain two geometrical measures of the overall curvature of the universe, one using the frequency of gravitational lenses, and the other using the characteristic pattern of the large-scale structure as a cosmic ruler.

Negative pressure is believed to be a property of vacuum energy, but the exact nature and existence of dark energy remains one of the great mysteries of the Big Bang. Results from the WMAP team in 2008 are in accordance with a universe that consists of 73% dark energy, 23% dark matter, 4.6% regular matter and less than 1% neutrinos. According to theory, the energy density in matter decreases with the expansion of the universe, but the dark energy density remains constant (or nearly so) as the universe expands. Therefore, matter made up a larger fraction of the total energy of the universe in the past than it does today, but its fractional contribution will fall in the far future as dark energy becomes even more dominant.

The dark energy component of the universe has been explained by theorists using a variety of competing theories including Einstein's cosmological constant but also extending to more exotic forms of quintessence or other modified gravity schemes. A cosmological constant problem, sometimes called the "most embarrassing problem in physics", results from the apparent discrepancy between the measured energy density of dark energy, and the one naively predicted from Planck units.

Dark Matter

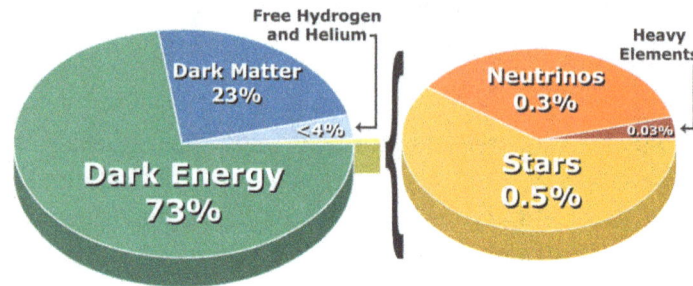

Chart shows the proportion of different components of the universe – about 95% is dark matter and dark energy.

During the 1970s and the 1980s, various observations showed that there is not sufficient visible matter in the universe to account for the apparent strength of gravitational forces within and between galaxies. This led to the idea that up to 90% of the matter in the universe is dark matter that does not emit light or interact with normal baryonic matter. In addition, the assumption that the universe is mostly normal matter led to predictions that were strongly inconsistent with observations. In particular, the universe today is far more lumpy and contains far less deuterium than can be accounted for without dark matter. While dark matter has always been controversial, it is inferred by various observations: the anisotropies in the CMB, galaxy cluster velocity dispersions, large-scale structure distributions, gravitational lensing studies, and X-ray measurements of galaxy clusters.

Indirect evidence for dark matter comes from its gravitational influence on other matter, as no dark matter particles have been observed in laboratories. Many particle physics candidates for dark matter have been proposed, and several projects to detect them directly are underway.

Additionally, there are outstanding problems associated with the currently favored cold dark matter model which include the dwarf galaxy problem and the cuspy halo problem. Alternative theories have been proposed that do not require a large amount of undetected matter, but instead modify the laws of gravity established by Newton and Einstein; yet no alternative theory has been as successful as the cold dark matter proposal in explaining all extant observations.

Horizon Problem

The horizon problem results from the premise that information cannot travel faster than light. In a universe of finite age this sets a limit—the particle horizon—on the separation of any two regions of space that are in causal contact. The observed isotropy of the CMB is problematic in this regard: if the universe had been dominated by radiation or matter at all times up to the epoch of last scattering, the particle horizon at that time would correspond to about 2 degrees on the sky. There would then be no mechanism to cause wider regions to have the same temperature.

A resolution to this apparent inconsistency is offered by inflationary theory in which a homogeneous and isotropic scalar energy field dominates the universe at some very early period (before baryogenesis). During inflation, the universe undergoes exponential expansion, and the particle horizon expands much more rapidly than previously assumed, so that regions presently on opposite sides of the observable universe are well inside each other's particle horizon. The observed isotropy of the CMB then follows from the fact that this larger region was in causal contact before the beginning of inflation.

Heisenberg's uncertainty principle predicts that during the inflationary phase there would be quantum thermal fluctuations, which would be magnified to cosmic scale. These fluctuations serve as the seeds of all current structure in the universe. Inflation predicts that the primordial fluctuations are nearly scale invariant and Gaussian, which has been accurately confirmed by measurements of the CMB.

If inflation occurred, exponential expansion would push large regions of space well beyond our observable horizon.

A related issue to the classic horizon problem arises because in most standard cosmological inflation models, inflation ceases well before electroweak symmetry breaking occurs, so inflation should not be able to prevent large-scale discontinuities in the electroweak vacuum since distant parts of the observable universe were causally separate when the electroweak epoch ended.

Magnetic Monopoles

The magnetic monopole objection was raised in the late 1970s. Grand unified theories predicted topological defects in space that would manifest as magnetic monopoles. These objects would be produced efficiently in the hot early universe, resulting in a density much higher than is consistent with observations, given that no monopoles have been found. This problem is also resolved by cosmic inflation, which removes all point defects from the observable universe, in the same way that it drives the geometry to flatness.

Flatness Problem

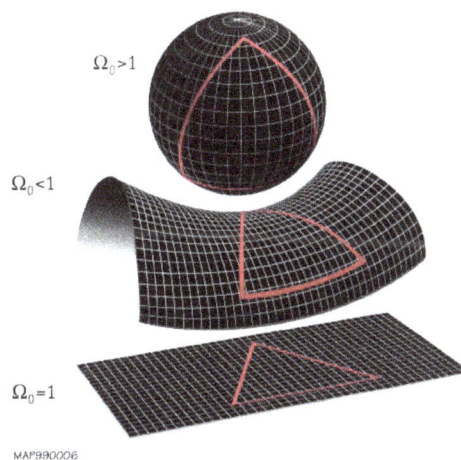

The overall geometry of the universe is determined by whether the Omega cosmological parameter is less than, equal to or greater than 1. Shown from top to bottom are a closed universe with positive curvature, a hyperbolic universe with negative curvature and a flat universe with zero curvature.

The flatness problem (also known as the oldness problem) is an observational problem associated with a Friedmann–Lemaître–Robertson–Walker metric (FLRW). The universe may have positive, negative, or zero spatial curvature depending on its total energy density. Curvature is negative, if its density is less than the critical density; positive, if greater; and zero at the critical density, in which case space is said to be *flat*.

The problem is that any small departure from the critical density grows with time, and yet the universe today remains very close to flat. Given that a natural timescale for departure from flatness might be the Planck time, 10^{-43} seconds, the fact that the universe has reached neither a heat death nor a Big Crunch after billions of years requires an explanation. For instance, even at the relatively late age of a few minutes (the time of nucleosynthesis), the density of the universe must have been within one part in 10^{14} of its critical value, or it would not exist as it does today.

Cause

Gottfried Wilhelm Leibniz wrote: *"Why is there something rather than nothing? The sufficient reason [...] is found in a substance which [...] is a necessary being bearing the reason for its existence within itself."* Philosopher of physics Dean Rickles has argued that numbers and mathematics (or their underlying laws) may necessarily exist. Physics may conclude that time did not exist before 'Big Bang', but 'started' with the Big Bang and hence there might be no 'beginning', 'before' or potentially 'cause' and instead always existed. Some also argue that nothing cannot exist or that non-existence might never have been an option. Quantum fluctuations, or other laws of physics that may have existed at the start of the Big Bang could then create the conditions for matter to occur.

Ultimate Fate of the Universe

Before observations of dark energy, cosmologists considered two scenarios for the future of the universe. If the mass density of the universe were greater than the critical density, then the universe would reach a maximum size and then begin to collapse. It would become denser and hotter again, ending with a state similar to that in which it started—a Big Crunch.

Alternatively, if the density in the universe were equal to or below the critical density, the expansion would slow down but never stop. Star formation would cease with the consumption of interstellar gas in each galaxy; stars would burn out, leaving white dwarfs, neutron stars, and black holes. Very gradually, collisions between these would result in mass accumulating into larger and larger black holes. The average temperature of the universe would asymptotically approach absolute zero—a Big Freeze. Moreover, if the proton were unstable, then baryonic matter would disappear, leaving only radiation and black holes. Eventually, black holes would evaporate by emitting Hawking radiation. The entropy of the universe would increase to the point where no organized form of energy could be extracted from it, a scenario known as heat death.

Modern observations of accelerating expansion imply that more and more of the currently visible universe will pass beyond our event horizon and out of contact with us. The eventual result is not known. The ΛCDM model of the universe contains dark energy in the form of a cosmological constant. This theory suggests that only gravitationally bound systems, such as galaxies, will remain together, and they too will be subject to heat death as the universe expands and cools. Other explanations of dark energy, called phantom energy theories, suggest that ultimately galaxy clusters, stars, planets, atoms, nuclei, and matter itself will be torn apart by the ever-increasing expansion in a so-called Big Rip.

Misconceptions

The following is a partial list of the popular misconceptions about the Big Bang model:

The Big Bang as the origin of the universe: One of the common misconceptions about the Big Bang model is the belief that it was the origin of the universe. However, the Big Bang model does not comment about how the universe came into being. Current conception of the Big Bang model assumes the existence of energy, time, and space, and does not comment about their origin or the cause of the dense and high temperature initial state of the universe.

The Big Bang was "small": It is misleading to visualize the Big Bang by comparing its size to everyday objects. When the size of the universe at Big Bang is described, it refers to the size of the observable universe, and not the entire universe.

Hubble's law violates the special theory of relativity: Hubble's law predicts that galaxies that are beyond Hubble Distance recede faster than the speed of light. However, special relativity does not apply beyond motion through space. Hubble's law describes velocity that results from expansion *of* space, rather than *through* space.

Doppler redshift vs cosmological red-shift: Astronomers often refer to the cosmological red-shift as a normal Doppler shift, which is a misconception. Although similar, the cosmological red-shift is not identical to the Doppler redshift. The Doppler redshift is based on special relativity, which does not consider the expansion of space. On the contrary, the cosmological red-shift is based on general relativity, in which the expansion of space is considered. Although they may appear identical for nearby galaxies, it may cause confusion if the behavior of distant galaxies is understood through the Doppler redshift.

Speculations

While the Big Bang model is well established in cosmology, it is likely to be refined. The Big Bang theory, built upon the equations of classical general relativity, indicates a singularity at the origin of cosmic time; this infinite energy density is regarded as impossible in physics. Still, it is known that the equations are not applicable before the time when the universe cooled down to the Planck temperature, and this conclusion depends on various assumptions, of which some could never be experimentally verified.

One proposed refinement to avoid this would-be singularity is to develop a correct treatment of quantum gravity.

It is not known what could have preceded the hot dense state of the early universe or how and why it originated, though speculation abounds in the field of cosmogony.

Some proposals, each of which entails untested hypotheses, are:

- Models including the Hartle–Hawking no-boundary condition, in which the whole of space-time is finite; the Big Bang does represent the limit of time but without any singularity.

- Big Bang lattice model, states that the universe at the moment of the Big Bang consists of an infinite lattice of fermions, which is smeared over the fundamental domain so it has rotational, translational and gauge symmetry. The symmetry is the largest symmetry possible and hence the lowest entropy of any state.

- Brane cosmology models, in which inflation is due to the movement of branes in string theory; the pre-Big Bang model; the ekpyrotic model, in which the Big Bang is the result of a collision between branes; and the cyclic model, a variant of the ekpyrotic model in which collisions occur periodically. In the latter model the Big Bang was preceded by a Big Crunch and the universe cycles from one process to the other.

- Eternal inflation, in which universal inflation ends locally here and there in a random fashion, each end-point leading to a *bubble universe*, expanding from its own big bang.

Proposals in the last two categories, see the Big Bang as an event in either a much larger and older universe or in a multiverse.

Religious and Philosophical Interpretations

As a description of the origin of the universe, the Big Bang has significant bearing on religion and philosophy. As a result, it has become one of the liveliest areas in the discourse between science and religion. Some believe the Big Bang implies a creator, and some see its mention in their holy books, while others argue that Big Bang cosmology makes the notion of a creator superfluous.

Big Bang Nucleosynthesis

In physical cosmology, Big Bang nucleosynthesis (or primordial nucleosynthesis) refers to the production of nuclei other than H-1, the normal, light hydrogen, during the early phases of the universe, shortly after the Big Bang.

It is believed to be responsible for the formation of hydrogen (H-1 or simply H), its isotope deuterium (H-2 or D), the helium isotopes He-3 and He-4, and the lithium isotope Li-7.

Big Bang nucleosynthesis begins about one minute after the Big Bang, when the universe has cooled enough to form stable protons and neutrons, after baryogenesis.

From simple thermodynamical arguments, one can calculate the fraction of protons and neutrons based on the temperature at this point.

This fraction is in favour of protons, because the higher mass of the neutron results in a spontaneous decay of neutrons to protons with a half-life of about 15 minutes.

Theory and Observation

Big Bang Nucleosynthesis was incapable to produce heavier atomic nuclei such as those necessary to build human bodies or a planet like the earth. Instead, those nuclei were formed in the interior of stars. By the same token, the element abundances we see around us are not the "primordial abundances" right after Big Bang Nucleosynthesis, but have been altered by later stellar processing.

While, in observing far-away objects, we always look back in time, it is impossible to look back directly to the time of Big Bang Nucleosynthesis since until a much later cosmic time of 400,000 years, the early universe was completely opaque. Instead, astronomers need to look for objects in the universe in which, to the best of current knowledge, the abundances of various elements are as close to their primordial value as possible, or which allow extrapolation to the primordial value. The nature of these estimates is described in the spotlight text Elements of the past: Reconstructing the original abundances of light nuclei.

To confront theory and observation, it is customary to plot the predictions against a parameter denoted by the greek letter eta, which is defined as the total number of protons and neutrons in our universe, divided by the number of photons in the cosmic background radiation. This ratio is nearly constant over time, and it is directly related to the density of nuclear building blocks in the early universe - an important ingredient for the Big Bang Nucleosynthesis calculations. An overview of the results is shown in the following diagram.

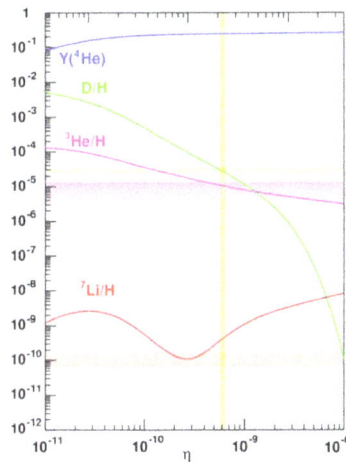

On the horizontal axis, we have the parameter eta. The scale is logarithmic; 10^{-9} corresponds to one proton or neutron for every billion photons, 10^{-8} to one for every hundred million, and so on (cf. the entry on exponential notation). The vertical axis indicates the different abundances. For for helium-4, it shows the mass ratio Y (the mass of helium-4 nuclei divided by the total mass of all protons and neutrons in the universe). For the other nuclei, it shows the number of such nuclei, divided by the number nuclei of hydrogen, the most abundant element. The curves indicate the theoretical predictions from Big Bang nucleosynthesis, the horizontal stripes the values that follow from observations.

While the abundance predictions have traditionally been used to fix the correct value for eta, there are different possibilities for measuring that number. Most notably, the presence of particles like protons and neutrons in the early universe leaves a slight, but measurable imprint on the cosmic background radiation. The value of eta that follows from recent high precision measurements with the Wilkinson Microwave Anisotropy Probe (WMAP) is indicated by the vertical golden strip.

As the diagram shows, within all theoretical and observational errors, the Big Bang Nucleosynthesis are impressively accurate. Except in the leftmost part, the pale blue strip indicating the observed value for helium-4 can hardly be distinguished from the theoretical curve. The green deuterium curve meets the pale green strip indicating the observed value almost exactly at the value indicated by the WMAP observations (vertical golden strip), and similarly the WMAP observations, the magenta helium-3 curve and the observed upper limit for helium-3 coincide very well. Only for lithium-7 is there an appreciable gap between prediction and observation though, given the uncertainties of determining the initial abundance of this element from observations, this discrepancy is likely to teach us more about stellar physics than about Big Bang Nucleosynthesis. All in all, this match between theory and observation constitutes one of the big successes of the standard models of cosmology.

Characteristics

There are several important characteristics of Big Bang nucleosynthesis (BBN):

- The initial conditions (neutron-proton ratio) were set in the first second after the Big Bang.

- The universe was very close to homogeneous at this time, and strongly radiation-dominated.

- The fusion of nuclei occurred between roughly 10 seconds to 20 minutes after the Big Bang; this corresponds to the temperature range when the universe was cool enough for deuterium to survive, but hot and dense enough for fusion reactions to occur at a significant rate.

- It was widespread, encompassing the entire observable universe.

The key parameter which allows one to calculate the effects of BBN is the baryon/photon number ratio, which is a small number of order 6×10^{-10}. This parameter corresponds to the baryon density and controls the rate at which nucleons collide and react; from this it is possible to calculate element abundances after nucleosynthesis ends. Although the baryon per photon ratio is important in determining element abundances, the precise value makes little difference to the overall picture. Without major changes to the Big Bang theory itself, BBN will result in mass abundances of about 75% of hydrogen-1, about 25% helium-4, about 0.01% of deuterium and helium-3, trace amounts (on the order of 10^{-10}) of lithium, and negligible heavier elements. That the observed abundances in the universe are generally consistent with these abundance numbers is considered strong evidence for the Big Bang theory.

In this field, for historical reasons it is customary to quote the helium-4 fraction *by mass*, symbol Y, so that 25% helium-4 means that helium-4 atoms account for 25% of the mass, but less than 8% of the nuclei would be helium-4 nuclei. Other (trace) nuclei are usually expressed as number ratios to hydrogen.

Important Parameters

The creation of light elements during BBN was dependent on a number of parameters; among those was the neutron-proton ratio (calculable from Standard Model physics) and the baryon-photon ratio.

Neutron–proton Ratio

The neutron-proton ratio was set by Standard Model physics before the nucleosynthesis era, essentially within the first 1-second after the Big Bang. Neutrons can react with positrons or electron neutrinos to create protons and other products in one of the following reactions:

$$n^+ e^+ \rightleftharpoons \bar{v}_e + p$$

$$n + \nu_e \rightleftharpoons p + e^-$$

At times much earlier than 1 sec, these reactions were fast and maintained the n/p ratio close to 1:1. As the temperature dropped, the equilibrium shifted in favour of protons due to their slightly lower mass, and the n/p ratio smoothly decreased. These reactions continued until the decreasing temperature and density caused the reactions to become too slow, which occurred at about T = 0.7 MeV (time around 1 second) and is called the freeze out temperature. At freeze out, the neutron-proton ratio was about 1/6. However, free neutrons are unstable with a mean life of 880 sec; some neutrons decayed in the next few minutes before fusing into any nucleus, so the ratio of total neutrons to protons after nucleosynthesis ends is about 1/7. Almost all neutrons that fused instead of decaying ended up combined into helium-4, due to the fact that helium-4 has the highest binding energy per nucleon among light elements. This predicts that about 8% of all atoms should be helium-4, leading to a mass fraction of helium-4 of about 25%, which is in line with observations. Small traces of deuterium and helium-3 remained as there was insufficient time and density for them to react and form helium-4.

Baryon–photon Ratio

The baryon–photon ratio, η, is the key parameter determining the abundances of light elements after nucleosynthesis ends. Baryons and light elements can fuse in the following main reactions:

$$p + n \rightarrow {}^2\text{H} + \gamma$$
$$p + {}^2\text{H} \rightarrow {}^3\text{He} + \gamma$$
$${}^2\text{H} + {}^2\text{H} \rightarrow {}^3\text{He} + n$$
$${}^2\text{H} + {}^2\text{H} \rightarrow {}^3\text{He} + p$$
$${}^3\text{H} + {}^2\text{H} \rightarrow {}^4\text{He} + p$$
$${}^3\text{H} + {}^2\text{H} \rightarrow {}^4\text{He} + n$$

along with some other low-probability reactions leading to ⁷Li or ⁷Be. (An important feature is that there are no stable nuclei with mass 5 or 8, which implies that reactions adding one baryon to ⁴He, or fusing two ⁴He, do not occur). Most fusion chains during BBN ultimately terminate in ⁴He (helium-4), while "incomplete" reaction chains lead to small amounts of left-over ²H or ³He; the amount of these decreases with increasing baryon-photon ratio. That is, the larger the baryon-photon ratio the more reactions there will be and the more efficiently deuterium will be eventually transformed into helium-4. This result makes deuterium a very useful tool in measuring the baryon-to-photon ratio.

Sequence

Big Bang nucleosynthesis began roughly 10 seconds after the big bang, when the universe had cooled sufficiently to allow deuterium nuclei to survive disruption by high-energy photons. This time is essentially independent of dark matter content, since the universe was highly radiation dominated until much later, and this dominant component controls the temperature/time relation.

At this time there were about six protons for every neutron, but a small fraction of the neutrons decay before fusing in the next few hundred seconds, so at the end of nucleosynthesis there are about seven protons to every neutron, and almost all the neutrons are in Helium-4 nuclei. The sequence of these reaction chains is shown on the image.

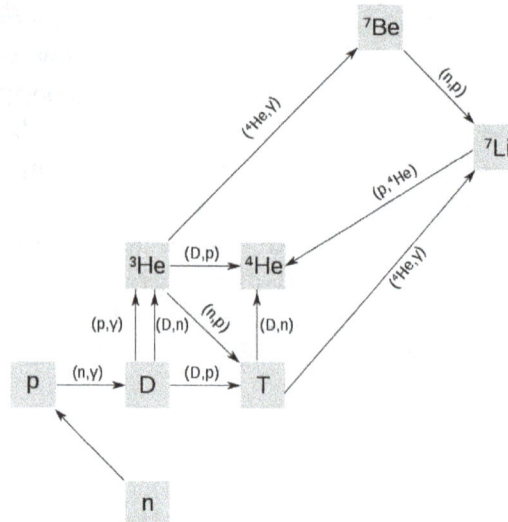

The main nuclear reaction chains for Big Bang nucleosynthesis

One feature of BBN is that the physical laws and constants that govern the behavior of matter at these energies are very well understood, and hence BBN lacks some of the speculative uncertainties that characterize earlier periods in the life of the universe. Another feature is that the process of nucleosynthesis is determined by conditions at the start of this phase of the life of the universe, and proceeds independently of what happened before.

As the universe expands, it cools. Free neutrons are less stable than helium nuclei, and the protons and neutrons have a strong tendency to form helium-4. However, forming helium-4 requires the intermediate step of forming deuterium. Before nucleosynthesis began, the temperature was high enough for many photons to have energy greater than the binding energy of deuterium; therefore any deuterium that was formed was immediately destroyed (a situation known as the deuterium bottleneck). Hence, the formation of helium-4 is delayed until the universe became cool enough for deuterium to survive (at about T = 0.1 MeV); after which there was a sudden burst of element formation. However, very shortly thereafter, around twenty minutes after the Big Bang, the temperature and density became too low for any significant fusion to occur. At this point, the elemental abundances were nearly fixed, and the only changes were the result of the radioactive decay of the two major unstable products of BBN, tritium and beryllium-7.

Theory

The history of Big Bang nucleosynthesis began with the calculations of Ralph Alpher in the 1940s. Alpher published the Alpher–Bethe–Gamow paper that outlined the theory of light-element production in the early universe.

During the 1970s, there was a major puzzle in that the density of baryons as calculated by Big Bang nucleosynthesis was much less than the observed mass of the universe based on measurements of galaxy rotation curves and galaxy cluster dynamics. This puzzle was resolved in large part by postulating the existence of dark matter.

Heavy Elements

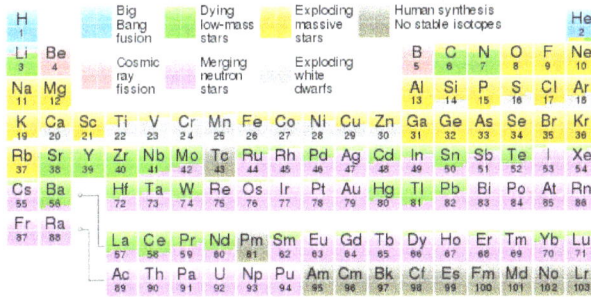

A version of the periodic table indicating the origins – including big bang nucleosynthesis – of the elements. All elements above 103 (lawrencium) are also manmade and are not included.

Big Bang nucleosynthesis produced very few nuclei of elements heavier than lithium due to a bottleneck: the absence of a stable nucleus with 8 or 5 nucleons. This deficit of larger atoms also limited the amounts of lithium-7 produced during BBN. In stars, the bottleneck is passed by triple collisions of helium-4 nuclei, producing carbon (the triple-alpha process). However, this process is very slow and requires much higher densities, taking tens of thousands of years to convert a significant amount of helium to carbon in stars, and therefore it made a negligible contribution in the minutes following the Big Bang.

The predicted abundance of CNO isotopes produced in Big Bang nucleosynthesis is expected to be on the order of 10^{-15} that of H, making them essentially undetectable and negligible. Indeed, none of these primordial isotopes of the elements from lithium to oxygen have yet been detected, although those of beryllium and boron may be able to be detected in the future. So far, the only stable nuclides known experimentally to have been made before or during Big Bang nucleosynthesis are protium, deuterium, helium-3, helium-4, and lithium-7.

Helium-4

Big Bang nucleosynthesis predicts a primordial abundance of about 25% helium-4 by mass, irrespective of the initial conditions of the universe. As long as the universe was hot enough for protons and neutrons to transform into each other easily, their ratio, determined solely by their relative masses, was about 1 neutron to 7 protons (allowing for some decay of neutrons into protons). Once it was cool enough, the neutrons quickly bound with an equal number of protons to form first deuterium, then helium-4. Helium-4 is very stable and is nearly the end of this chain if it runs for only a short time, since helium neither decays nor combines easily to form heavier nuclei (since there are no stable nuclei with mass numbers of 5 or 8, helium does not combine easily with either protons, or with itself). Once temperatures are lowered, out of every 16 nucleons (2 neutrons and 14 protons), 4 of these (25% of the total particles and total mass) combine quickly into one helium-4 nucleus. This produces one helium for every 12 hydrogens, resulting in a universe that is a little over 8% helium by number of atoms, and 25% helium by mass.

One analogy is to think of helium-4 as ash, and the amount of ash that one forms when one completely burns a piece of wood is insensitive to how one burns it. The resort to the BBN theory of the helium-4 abundance is necessary as there is far more helium-4 in the universe than can be explained by stellar nucleosynthesis. In addition, it provides an important test for the Big Bang theory. If the observed helium abundance is significantly different from 25%, then this would pose a serious challenge to the theory. This would particularly be the case if the early helium-4 abundance was much smaller than 25% because it is hard to destroy helium-4. For a few years during the mid-1990s, observations suggested that this might be the case, causing astrophysicists to talk about a Big Bang nucleosynthetic crisis, but further observations were consistent with the Big Bang theory.

Deuterium

Deuterium is in some ways the opposite of helium-4, in that while helium-4 is very stable and difficult to destroy, deuterium is only marginally stable and easy to destroy. The temperatures, time, and densities were sufficient to combine a substantial fraction of the deuterium nuclei to form helium-4 but insufficient to carry the process further using helium-4 in the next fusion step. BBN did not convert all of the deuterium in the universe to helium-4 due to the expansion that cooled the universe and reduced the density, and so cut that conversion short before it could proceed any further. One consequence of this is that, unlike helium-4, the amount of deuterium is very sensitive to initial conditions. The denser the initial universe was, the more deuterium would be converted to helium-4 before time ran out, and the less deuterium would remain.

There are no known post-Big Bang processes which can produce significant amounts of deuterium. Hence observations about deuterium abundance suggest that the universe is not infinitely old, which is in accordance with the Big Bang theory.

During the 1970s, there were major efforts to find processes that could produce deuterium, but those revealed ways of producing isotopes other than deuterium. The problem was that while the concentration of deuterium in the universe is consistent with the Big Bang model as a whole, it is too high to be consistent with a model that presumes that most of the universe is composed of protons and neutrons. If one assumes that all of the universe consists of protons and neutrons, the density of the universe is such that much of the currently observed deuterium would have been burned into helium-4. The standard explanation now used for the abundance of deuterium is that the universe does not consist mostly of baryons, but that non-baryonic matter (also known as dark matter) makes up most of the mass of the universe. This explanation is also consistent with calculations that show that a universe made mostly of protons and neutrons would be far more *clumpy* than is observed.

It is very hard to come up with another process that would produce deuterium other than by nuclear fusion. Such a process would require that the temperature be hot enough to produce deuterium, but not hot enough to produce helium-4, and that this process should immediately cool to non-nuclear temperatures after no more than a few minutes. It would also be necessary for the deuterium to be swept away before it reoccurs.

Producing deuterium by fission is also difficult. The problem here again is that deuterium is very unlikely due to nuclear processes, and that collisions between atomic nuclei are likely to result either in the fusion of the nuclei, or in the release of free neutrons or alpha particles. During the

1970s, cosmic ray spallation was proposed as a source of deuterium. That theory failed to account for the abundance of deuterium, but led to explanations of the source of other light elements.

Measurements and Status of Theory

The theory of BBN gives a detailed mathematical description of the production of the light "elements" deuterium, helium-3, helium-4, and lithium-7. Specifically, the theory yields precise quantitative predictions for the mixture of these elements, that is, the primordial abundances at the end of the big-bang.

In order to test these predictions, it is necessary to reconstruct the primordial abundances as faithfully as possible, for instance by observing astronomical objects in which very little stellar nucleosynthesis has taken place (such as certain dwarf galaxies) or by observing objects that are very far away, and thus can be seen in a very early stage of their evolution (such as distant quasars).

As noted above, in the standard picture of BBN, all of the light element abundances depend on the amount of ordinary matter (baryons) relative to radiation (photons). Since the universe is presumed to be homogeneous, it has one unique value of the baryon-to-photon ratio. For a long time, this meant that to test BBN theory against observations one had to ask: can *all* of the light element observations be explained with a *single value* of the baryon-to-photon ratio? Or more precisely, allowing for the finite precision of both the predictions and the observations, one asks: is there some *range* of baryon-to-photon values which can account for all of the observations?

More recently, the question has changed: Precision observations of the cosmic microwave background radiation with the Wilkinson Microwave Anisotropy Probe (WMAP) and Planck give an independent value for the baryon-to-photon ratio. Using this value, are the BBN predictions for the abundances of light elements in agreement with the observations?

The present measurement of helium-4 indicates good agreement, and yet better agreement for helium-3. But for lithium-7, there is a significant discrepancy between BBN and WMAP/Planck, and the abundance derived from Population II stars. The discrepancy is a factor of $2.4-4.3$ below the theoretically predicted value and is considered a problem for the original models, that have resulted in revised calculations of the standard BBN based on new nuclear data, and to various reevaluation proposals for primordial proton-proton nuclear reactions, especially the abundances of $^7Be + n \rightarrow ^7Li + p$, versus $^7Be + ^2H \rightarrow ^8Be + p$.

Non-standard Scenarios

In addition to the standard BBN scenario there are numerous non-standard BBN scenarios. These should not be confused with non-standard cosmology: a non-standard BBN scenario assumes that the Big Bang occurred, but inserts additional physics in order to see how this affects elemental abundances. These pieces of additional physics include relaxing or removing the assumption of homogeneity, or inserting new particles such as massive neutrinos.

There have been, and continue to be, various reasons for researching non-standard BBN. The first, which is largely of historical interest, is to resolve inconsistencies between BBN predictions and observations. This has proved to be of limited usefulness in that the inconsistencies were resolved by better observations, and in most cases trying to change BBN resulted in abundances that were more inconsistent with observations rather than less. The second reason for researching non-standard BBN, and largely the focus of non-standard BBN in the early 21st century, is to use BBN to place limits on unknown or speculative physics. For example, standard BBN assumes that no exotic hypothetical particles were involved in BBN. One can insert a hypothetical particle (such as a massive neutrino) and see what has to happen before BBN predicts abundances that are very different from observations. This has been done to put limits on the mass of a stable tau neutrino.

Big Bang Nucleosynthesis

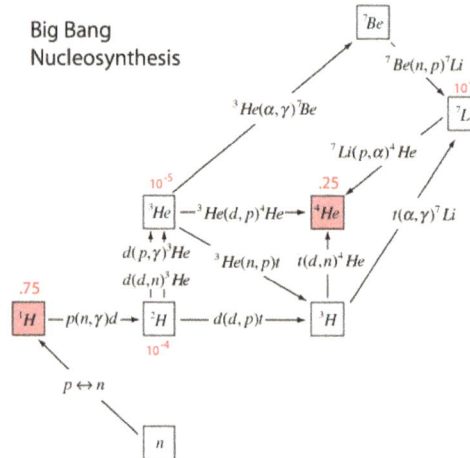

The modeling of the early universe by the standard big bang model gives a scenario that involves twelve nuclear interactions that led to the present cosmic abundances of elements outside the stars. The vast majority of the mass of ordinary matter in the universe is hydrogen and helium, remaining from the early stages of the universe. The illustration above gives nominal abundances of the current constituents. These abundances are model dependent and subject to revision.

Beryllium-7 is an isotope that is produced in this process, but is not a part of the cosmic abundances because it is radioactive with half-life 53.28 days, decaying to ^{7}Li. Neutrons are not part of the background elements because free neutrons decay with half-life 10.3 minutes. The other constituent of the reactions is tritium, ^{3}H, which has a half-life of 12.32 years.

Lambda-CDM Model

The term 'concordance model' is used in cosmology to indicate the *currently accepted* and most commonly used cosmological model. It is important to identify a concordance model because the measurement of many astrophysical quantities (e.g. distance, radius, luminosity and surface brightness) depend upon the cosmological model used. Consequently, for ease of comparison if nothing else, the models assumed in different studies should at least be similar, if not identical.

Currently, the concordance model is the Lambda CDM model (which includes cold dark matter and a cosmological constant). In this model the Universe is 13.7 billion years old and made up of 4% baryonic matter, 23% dark matter and 73% dark energy. The Hubble constant for this model is 71 km/s/Mpc and the density of the Universe is very close to the critical value for re-collapse. These values were derived from WMAP (Wilkinson Microwave Anisotropy Probe) satellite observations of the cosmic microwave background radiation.

ΛCDM Model of Cosmology

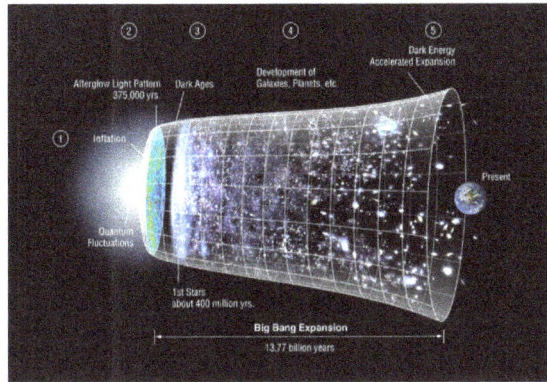

In this picture, the infant universe is an extremely hot, dense, nearly homogeneous mixture of photons and matter, tightly coupled together as a plasma. An approximate graphical timeline of its theoretical evolution is shown in the figure above, with numbers keyed to the explanatory text below.

1. The initial conditions of this early plasma are currently thought to be established during a period of rapid expansion known as inflation. Density fluctuations in the primordial plasma are seeded by quantum fluctuations in the field driving inflation. The amplitude of the primordial gravitational potential fluctuations is nearly the same on all spatial scales.

    The small perturbations propagate through the plasma collisionally as a sound wave, producing under- and overdensities in the plasma with simultaneous changes in density of matter and radiation. CDM doesn't share in these pressure-induced oscillations, but does act gravitationally, either enhancing or negating the acoustic pattern for the photons and baryons (Hu & White 2004).

2. Eventually physical conditions in the expanding, cooling plasma reach the point where electrons and baryons are able to stably recombine, forming atoms, mostly in the form of neutral hydrogen. The photons decouple from the baryons as the plasma becomes neutral, and perturbations no longer propagate as acoustic waves: the existing density pattern becomes "frozen". This snapshot of the density fluctuations is preserved in the CMB anisotropies and the imprint of baryon acoustic oscillations (BAO) observable today in large scale structure (Eisenstein & Hu 1998).

3. Recombination produces a largely neutral universe which is unobservable throughout most of the electromagnetic spectrum, an era sometimes referred to as the "Dark Ages". During this era, CDM begins gravitational collapse in overdense regions. Baryonic matter gravitationally collapses into these CDM halos, and "Cosmic Dawn" begins with the formation of the first radiation sources such as stars. Radiation from these objects reionizes the intergalactic medium.

4. Structure continues to grow and merge under the influence of gravity, forming a vast cosmic web of dark matter density. The abundance of luminous galaxies traces the statistics of the underlying matter density. Clusters of galaxies are the largest bound objects. Despite this reorganization, galaxies retain the BAO correlation length that was established in the era of the CMB.

5. As the universe continues to expand over time, the negative pressure associated with the cosmological constant (the form of dark energy in ΛCDM) increasingly dominates over opposing gravitational forces, and the expansion of the universe accelerates.

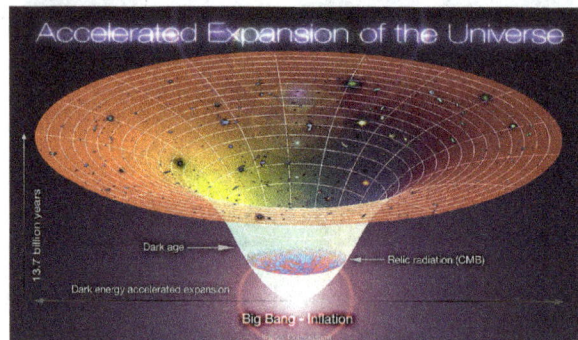

Lambda-CDM, accelerated expansion of the universe. The time-line in this schematic diagram extends from the Big Bang/inflation era 13.7 Byr ago to the present cosmological time.

Most modern cosmological models are based on the cosmological principle, which states that our observational location in the universe is not unusual or special; on a large-enough scale, the universe looks the same in all directions (isotropy) and from every location (homogeneity).

The model includes an expansion of metric space that is well documented both as the red shift of prominent spectral absorption or emission lines in the light from distant galaxies and as the time dilation in the light decay of supernova luminosity curves. Both effects are attributed to a Doppler shift in electromagnetic radiation as it travels across expanding space. Although this expansion increases the distance between objects that are not under shared gravitational influence, it does not increase the size of the objects (e.g. galaxies) in space. It also allows for distant galaxies to recede from each other at speeds greater than the speed of light; local expansion is less than the speed of light, but expansion summed across great distances can collectively exceed the speed of light.

The letter Λ (lambda) represents the cosmological constant, which is currently associated with a vacuum energy or dark energy in empty space that is used to explain the contemporary accelerating expansion of space against the attractive effects of gravity. A cosmological constant has negative pressure, $p = -\rho c^2$, , which contributes to the stress-energy tensor that, according to the general theory of relativity, causes accelerating expansion. The fraction of the total energy density of our (flat or almost flat) universe that is dark energy, Ω_Λ, is currently [2015] estimated to be 0.692 ± 0.012, or even 0.6911 ± 0.0062 based on Planck satellite data.

Dark matter is postulated in order to account for gravitational effects observed in very large-scale structures (the "flat" rotation curves of galaxies; the gravitational lensing of light by galaxy clusters; and enhanced clustering of galaxies) that cannot be accounted for by the quantity of observed matter. Cold dark matter is *non-baryonic*, i.e. it consists of matter other than protons and neutrons (and electrons, by convention, although electrons are not baryons); *cold*, i.e. its velocity is far less than the speed of light at the epoch of radiation-matter equality (thus neutrinos are excluded, being non-baryonic but not cold); *dissipationless*, i.e. it cannot cool by radiating photons; and *collisionless*, i.e. the dark matter particles interact with each other and other particles only through gravity and possibly the weak force. The dark matter component is currently [2013] estimated to constitute about 26.8% of the mass-energy density of the universe.

The remaining 4.9% [2013] comprises all ordinary matter observed as atoms, chemical elements, gas and plasma, the stuff of which visible planets, stars and galaxies are made. The great majority of ordinary matter in the universe is unseen, since visible stars and gas inside galaxies and clusters account for less than 10 per cent of the ordinary matter contribution to the mass-energy density of the universe.

Also, the energy density includes a very small fraction (~ 0.01%) in cosmic microwave background radiation, and not more than 0.5% in relic neutrinos. Although very small today, these were much more important in the distant past, dominating the matter at redshift > 3200.

The model includes a single originating event, the "Big Bang", which was not an explosion but the abrupt appearance of expanding space-time containing radiation at temperatures of around 10^{15} K. This was immediately (within 10^{-29} seconds) followed by an exponential expansion of space by a scale multiplier of 10^{27} or more, known as cosmic inflation. The early universe remained hot (above 10,000 K) for several hundred thousand years, a state that is detectable as a residual cosmic microwave background, or CMB, a very low energy radiation emanating from all parts of the sky. The "Big Bang" scenario, with cosmic inflation and standard particle physics, is the only current cosmological model consistent with the observed continuing expansion of space, the observed distribution of lighter elements in the universe (hydrogen, helium, and lithium), and the spatial texture of minute irregularities (anisotropies) in the CMB radiation. Cosmic inflation also addresses the "horizon problem" in the CMB; indeed, it seems likely that the universe is larger than the observable particle horizon.

The model uses the Friedmann–Lemaître–Robertson–Walker metric, the Friedmann equations and the cosmological equations of state to describe the observable universe from right after the inflationary epoch to present and future.

Cosmic Expansion

The expansion of the universe is parametrized by a dimensionless scale factor $a = a(t)$ (with time t counted from the birth of the universe), defined relative to the present day, so $a_0 = a(t_0) = 1$; the usual convention in cosmology is that subscript 0 denotes present-day values, so t_0 is the current age of the universe. The scale factor is related to the observed redshift z of the light emitted at time t_{em} by

$$\frac{1}{a(t_{em})} = 1 + z.$$

The expansion rate is described by the time-dependent Hubble parameter, $H(t)$, defined as

$$H(t) \equiv \frac{\dot{a}}{a},$$

where $\dot{a}$ is the time-derivative of the scale factor. The first Friedmann equation gives the expansion rate in terms of the matter+radiation density ρ, the curvature k, and the cosmological constant Λ,

$$H^2 = \left(\frac{\dot{a}}{a}\right)^2 = \frac{8\pi G}{3}\rho - \frac{kc^2}{a^2} + \frac{\Lambda c^2}{3}$$

where as usual c is the speed of light and G is the gravitational constant. A critical density ρ_{crit} is the present-day density, which gives zero curvature k, assuming the cosmological constant Λ is zero, regardless of its actual value. Substituting these conditions to the Friedmann equation gives

$$\rho_{crit} = \frac{3H_0^2}{8\pi G} = 1.878\ 47(23)\times 10^{-26}\ h^2\ \text{kg m}^{-3}$$

where $h \equiv H_0/(100\ \text{km s}^{-1}\ \text{Mpc}^{-1})$ is the reduced Hubble constant. If the cosmological constant were actually zero, the critical density would also mark the dividing line between eventual recollapse of the universe to a Big Crunch, or unlimited expansion. For the Lambda-CDM model with a positive cosmological constant (as observed), the universe is predicted to expand forever regardless of whether the total density is slightly above or below the critical density; though other outcomes are possible in extended models where the dark energy is not constant but actually time-dependent.

It is standard to define the present-day density parameter Ω_x for various species as the dimensionless ratio

$$\Omega_x \equiv \frac{\rho_x(t=t_0)}{\rho_{crit}} = \frac{8\pi G\rho_x(t=t_0)}{3H_0^2}$$

where the subscript x is one of "b" for baryons, "c" for cold dark matter, "rad" for radiation (photons plus relativistic neutrinos), and "DE" or "Λ" for dark energy.

Since the densities of various species scale as different powers of a, e.g. a^{-3} for matter etc., the Friedmann equation can be conveniently rewritten in terms of the various density parameters as

$$H(a) \equiv \frac{\dot{a}}{a} = H_0\sqrt{(\Omega_c + \Omega_b)a^{-3} + \Omega_{rad}a^{-4} + \Omega_k a^{-2} + \Omega_{DE}a^{-3(1+w)}}$$

where w is the equation of state of dark energy, and assuming negligible neutrino mass (significant neutrino mass requires a more complex equation). The various Ω parameters add up to 1 by construction. In the general case this is integrated by computer to give the expansion history a(t) and also observable distance-redshift relations for any chosen values of the cosmological parameters, which can then be compared with observations such as supernovae and baryon acoustic oscillations.

In the minimal 6-parameter Lambda-CDM model, it is assumed that curvature Ω_k is zero and $w = -1$, so this simplifies to

$$H(a) = H_0\sqrt{\Omega_m a^{-3} + \Omega_{rad}a^{-4} + \Omega_\Lambda}$$

Observations show that the radiation density is very small today, $\Omega_{rad} \sim 10^{-4}$; if this term is neglected the above has an analytic solution

$$a(t) = (\Omega_m / \Omega_\Lambda)^{1/3} \sinh^{2/3}(t / t_\Lambda)$$

where $t_\Lambda \equiv 2 / (3H_0\sqrt{\Omega_\Lambda})$; this is fairly accurate for a > 0.01 or t > 10 Myr. Solving for $a(t) = 1$ gives the present age of the universe t_0 in terms of the other parameters.

It follows that the transition from decelerating to accelerating expansion (the second derivative $\ddot{a}$ crossing zero) occurred when

$$a = (\Omega_m / 2\Omega_\Lambda)^{1/3}$$

which evaluates to a ~ 0.6 or z ~ 0.66 for the best-fit parameters estimated from the Planck space-craft.

Parameters

Planck Collaboration Cosmological parameters			
	Description	**Symbol**	**Value**
Independent parameters	Physical baryon density parameter	$\Omega_b h^2$	0.02230±0.00014
	Physical dark matter density parameter	$\Omega_c h^2$	0.1188±0.0010
	Age of the universe	t_0	13.799±0.021 × 10^9 years
	Scalar spectral index	n_s	0.9667±0.0040
	Curvature fluctuation amplitude, k_0 = 0.002 Mpc^{-1}	Δ^2_R	2.441+0.088 −0.092×10^{-9}
	Reionization optical depth	τ	0.066±0.012
Fixed parameters	Total density parameter	Ω_{tot}	1
	Equation of state of dark energy	w	−1
	Sum of three neutrino masses	Σm_ν	0.06 eV/c
	Effective number of relativistic degrees of freedom	N_{eff}	3.046
	Tensor/scalar ratio	r	0
	Running of spectral index	$d n_s / d \ln k$	0
Calculated values	Hubble constant	H_0	67.74±0.46 km s^{-1} Mpc^{-1}
	Baryon density parameter	Ω_b	0.0486±0.0010
	Dark matter density parameter	Ω_c	0.2589±0.0057
	Matter density parameter	Ω_m	0.3089±0.0062
	Dark energy density parameter	Ω_Λ	0.6911±0.0062
	Critical density	ρ_{crit}	(8.62±0.12)×10^{-27} kg/m³
	Fluctuation amplitude at 8h^{-1} Mpc	σ_8	0.8159±0.0086
	Redshift at decoupling	z_*	1089.90±0.23
	Age at decoupling	t_*	377700±3200 years
	Redshift of reionization (with uniform prior)	z_{re}	8.5+1.0 −1.1

The simple ΛCDM model is based on six parameters: physical baryon density parameter; physical dark matter density parameter; the age of the universe; scalar spectral index; curvature fluctuation amplitude; and reionization optical depth. In accordance with Occam's razor, six is the smallest number of parameters needed to give an acceptable fit to current observations; other possible parameters are fixed at "natural" values, e.g. total density parameter = 1.00, dark energy equation of state = −1.

The values of these six parameters are mostly not predicted by current theory (though, ideally, they may be related by a future "Theory of Everything"), except that most versions of cosmic inflation predict the scalar spectral index should be slightly smaller than 1, consistent with the estimated value 0.96. The parameter values, and uncertainties, are estimated using large computer searches to locate the region of parameter space providing an acceptable match to cosmological observations. From these six parameters, the other model values, such as the Hubble constant and the dark energy density, can be readily calculated.

Commonly, the set of observations fitted includes the cosmic microwave background anisotropy, the brightness/redshift relation for supernovae, and large-scale galaxy clustering including the baryon acoustic oscillation feature. Other observations, such as the Hubble constant, the abundance of galaxy clusters, weak gravitational lensing and globular cluster ages, are generally consistent with these, providing a check of the model, but are less precisely measured at present.

Parameter values listed below are from the Planck Collaboration Cosmological parameters 68% confidence limits for the base ΛCDM model from Planck CMB power spectra, in combination with lensing reconstruction and external data (BAO+JLA+H_0).

1. The "physical baryon density parameter" $\Omega_b h^2$ is the "baryon density parameter" Ω_b multiplied by the square of the reduced Hubble constant $h = H_0 / (100$ km s^{-1} Mpc$^{-1})$. Likewise for the difference between "physical dark matter density parameter" and "dark matter density parameter".

2. A density $\rho_{x = \Omega_x}\rho$crit is expressed in terms of the critical density ρcrit, which is the total density of matter/energy needed for the universe to be spatially flat. Measurements indicate that the actual total density ρ_{tot} is very close if not equal to this value.

3. This is the minimal value allowed by solar and terrestrial neutrino oscillation experiments.

4. from the Standard Model of particle physics

5. Calculated from Ω_{bh}^2 and $h = H_0 / (100$ km s^{-1} Mpc$^{-1})$.

6. Calculated from Ω_{ch}^2 and $h = H_0 / (100$ km s^{-1} Mpc$^{-1})$.

7. Calculated from $h = H_0 / (100$ km s^{-1} Mpc$^{-1})$ per $\rho_{crit} = 1.87847 \times 10^{-26} h^2$ kg m^{-3}.

Missing Baryon Problem

Massimo Persic and Paolo Salucci firstly estimated the baryonic density today present in ellipticals, spirals, groups and clusters of galaxies. They performed an integration of the baryonic mass-to-light ratio over luminosity (in the following M_b / L), weighted with the luminosity function $\phi(L)$ over the previously mentioned classes of astrophysical objects:

$$\rho_b = \sum \int L\phi(L)\frac{M_b}{L}\,dL.$$

The result was:

$$\Omega_b = \Omega_* + \Omega_{gas} = 2.2\times10^{-3} + 1.5\times10^{-3}\,h^{-1.3} \simeq 0.003$$

where $h \simeq 0.72$.

Note that this value is much lower than the prediction of standard cosmic nucleosynthesis $\Omega_b \simeq 0.048$, so that stars and gas in galaxies and in galaxy groups and clusters account for less than the 10 per cent of the primordially synthesized baryons. This issue is known as the problem of the "missing baryons".

Extended Models

Extended model parameters		
Description	**Symbol**	**Value**
Total density parameter	Ω_{tot}	1.0023+0.0056 −0.0054
Equation of state of dark energy	w	−0.980±0.053
Tensor-to-scalar ratio	r	< 0.11, k_0 = 0.002 Mpc^{-1} (2σ)
Running of the spectral index	$d\,n_s / d\ln k$	−0.022±0.020, k_0 = 0.002 Mpc^{-1}
Physical neutrino density parameter	$\Omega_\nu h^2$	< 0.0062
Sum of three neutrino masses	$\sum m_\nu$	< 0.58 eV/c^2 (2σ)

Extended models allow one or more of the "fixed" parameters above to vary, in addition to the basic six; so these models join smoothly to the basic six-parameter model in the limit that the additional parameter(s) approach the default values. For example, possible extensions of the simplest ΛCDM model allow for spatial curvature (Ω_{tot} may be different from 1); or quintessence rather than a cosmological constant where the equation of state of dark energy is allowed to differ from −1. Cosmic inflation predicts tensor fluctuations (gravitational waves). Their amplitude is parameterized by the tensor-to-scalar ratio (denoted r), which is determined by the unknown energy scale of inflation. Other modifications allow hot dark matter in the form of neutrinos more massive than the minimal value, or a running spectral index; the latter is generally not favoured by simple cosmic inflation models.

Allowing additional variable parameter(s) will generally *increase* the uncertainties in the standard six parameters quoted above, and may also shift the central values slightly. The table shows results for each of the possible "6+1" scenarios with one additional variable parameter; this indicates that, as of 2015, there is no convincing evidence that any additional parameter is different from its default value.

Some researchers have suggested that there is a running spectral index, but no statistically significant study has revealed one. Theoretical expectations suggest that the tensor-to-scalar ratio r should be between 0 and 0.3, and the latest results are now within those limits.

Cosmological Constant

In the context of cosmology the cosmological constant is a homogeneous energy density that causes the expansion of the universe to accelerate. Originally proposed early in the development of general relativity in order to allow a static universe solution it was subsequently abandoned when the universe was found to be expanding. Now the cosmological constant is invoked to explain the observed *acceleration* of the expansion of the universe. The cosmological constant is the simplest realization of dark energy, which is the more generic name given to the unknown cause of the acceleration of the universe. Its existence is also predicted by quantum physics, where it enters as a form of vacuum energy, although the magnitude predicted by quantum theory does not match that observed in cosmology.

The Physics of the Cosmological Constant

To explore more deeply the nature of the Universe, we must use the mathematical language in Einstein's general relativity to relate the geometry of space-time (expressed by the metric tensor, $g\mu v$) to the energy content of the universe, (expressed by the energy-momentum tensor, $T\mu v$).

Einstein Field Equations

Arguably, one of Einstein's most significant discoveries was that the distribution of energy determines the geometry of space-time, which is encoded in his field equation,

$$R_{\mu v} - \frac{1}{2} R g_{\mu v} = 8\pi G T_{\mu v},$$

where G is the gravitational constant. Although this is the simplest form of the equations the freedom remains to add a constant term. This "cosmological constant" was what Einstein added in order to achieve a static universe, and it is given the symbol Λ.

$$R_{\mu v} - \frac{1}{2} R g_{\mu v} + \Lambda g_{\mu v} = 8\pi G T_{\mu v}$$

When Λ is positive it acts as a repulsive force.

Vacuum Energy

Vacuum energy arises naturally in quantum mechanics due to the uncertainty principle. In particle physics the vacuum refers to the ground state of the theory -- the lowest energy configuration. The uncertainty principle does not allow states of exactly zero energy, even in vacuum (virtual particles are created). Since in general relativity all forms of energy gravitate, this ground state vacuum energy impacts the dynamics of the expansion of the universe.

Equation of State (w)	
Radiation	1/3
Matter (pressureless)	0
Curvature	-1/3
Cosmological Constant	-1
Matter (general)	0<w<1/3
Quintessence	-1<w<-1/3

Figure: The equation of state, w , describes the relationship between pressure and density in a material according to $w=p/\rho$. Here are some examples of the equation of state for common fluids. When matter is at rest (pressureless dust) it has $w=0$, but as it picks up velocity (v) its equation of state increases until $w\rightarrow1/3$ as $v\rightarrow c$.

Vacuum energy should not have any dissipative processes such as heat conduction or viscosity, so it should take the form of a perfect fluid,

$$T_{\mu v} = (\rho + p)U_\mu U_v + pg_{\mu v}$$

In order to maintain Lorentz invariance, vacuum energy should also have no preferred direction. Therefore the first term in the perfect fluid energy tensor must be zero, requiring

$$p^{vac} = -\rho^{vac}$$

which corresponds to an equation of state $w^{vac}=p^{vac}/\rho^{vac}=-1$, and results in an energy-momentum tensor for vacuum energy,

$$T_{\mu v}^{vac} = p^{vac}g_{\mu v} = -\rho^{vac}g_{\mu v}$$

Equivalence of Cosmological Constant and Vacuum Energy

We can split the energy-momentum tensor into a term describing the matter and energy, and a term describing the vacuum $T_{\mu v} = T_{\mu v}^{matter} + T^{va}$, Einstein's equation including vacuum energy becomes,

$$R_{\mu v} - \frac{1}{2}Rg_{\mu v} = 8\pi G\ (T_{\mu v}^{matter} - \rho_{vac}g_{\mu v})$$

Recall that the cosmological constant enters Einstein's equation in the form,

$$R_{\mu v} - \frac{1}{2}Rg_{\mu v} + \Lambda g_{\mu v} = 8\pi GT_{\mu v}.$$

So vacuum energy and the cosmological constant have identical behaviour in general relativity, as long as the vacuum energy density is identified with,

$$p^{vac} = \frac{\Lambda}{8\pi G}.$$

Cosmology

Aleksandr Friedmann derived the famous equation which bears his name, shortly before
his untimely death aged just 37 from Typhoid fever. George Gamow was one of his students.

In an homogeneous, isotropic universe the geometry is defined by the Friedamnn-Lemaître-Robertson-Walker metric (FLRW metric) and the dynamics of the universe are governed by the Friedmann equations (Friedmann equations). The dynamics are driven by the energy content of the universe and the equation of state of the components that make up the energy density. The equation of state relates density ρ to pressure p according to $w=p/\rho$. The cosmological constant enters these equations in the following way, where a is the scale factor of the universe normalized to 1 at the present day, $H=a^{\cdot}/a$ is Hubble's constant (an overdot represents differentiation with respect to time), and k is the curvature of the universe given by +1, 0, and -1 for positive, flat, and negative curvature respectively,

$$H_2 = \frac{8\pi G}{3}\rho - \frac{k}{a^2} + \frac{\Lambda}{3}$$

$$\frac{\ddot{a}}{a} = -\frac{4\pi G}{3}(\rho+3p) + \frac{\Lambda}{3}.$$

These equations are more concisely written by considering both the cosmological constant and curvature as forms of energy density ($\rho_\Lambda = \Lambda/8\pi G$ and $\rho_k = -3k/8\pi Ga^2$). Then,

$$H^2 = \frac{8\pi G}{3}\sum_i \rho i,$$

$$\frac{\ddot{a}}{a} = -\frac{4\pi G}{3}\sum_i (\rho_i + 3p_i).$$

The different components have different equations of state, wi, which determines how their density changes with the expansion of the universe:

$$\rho_i = \rho_{i0} a^{-3(1+w_i)}$$

Pressureless matter has $w=0$, radiation has $w=1/3$, curvature has an effective $w=-1/3$, cosmological constant has $w=-1$.

The current energy density of each component, ρ_{io}, is often represented as a fraction of the critical density, $\rho_c = 3H_0^2/8\pi G$, which is the energy density required to close the universe (also calculated at the present day). Denoting this $\Omega i = \rho io/\rho c$ and using equation $\rho_i = \rho_{i0} a^{-3(1+w_i)}$ allows us to write

$$H^2 = H_0^2 \sum_i \frac{\rho_i}{\rho_c} = H_0^2 \sum_i \Omega_i a^{-3(1+w_i)}$$

For pressure to do work there needs to be a pressure gradient -- a relatively high pressure region next to a relatively low pressure region -- that will then cause movement from high pressure to low. In a homogeneous universe there are no pressure gradients, so a positive pressure does no work and has no expanding effect (there are no low-pressure regions for it to push matter into). On the contrary, in general relativity all forms of energy gravitate so pressure effectively pulls, strengthening the attractive force of gravity. The cosmological constant has negative pressure, $w=-1$, so its general relativistic contribution counteracts the normal force of gravity and provides an outwards acceleration.

Observational Evidence

Observational evidence for the accelerating universe is now very strong, with many different experiments covering vastly different timescales, length scales, and physical processes, all supporting the standard Λ CDM cosmological model, in which the universe is flat with an energy density made up of about 4% baryonic matter, 23% dark matter, and 73% cosmological constant.

The critical observational result that brought the cosmological constant into its modern prominence was the discovery that distant type Ia supernovae ($0<z<1$), used as standard candles, were *fainter* than expected in a decelerating universe (Riess et al. 1998, Perlmutter et al. 1999). Since then many groups have confirmed this result with more supernovae and over a larger range of redshifts. Of particular importance are the observations that extremely high redshift ($z>1$) supernovae are *brighter* than expected, which is the observational signature that is expected from a period of deceleration preceding our current period of acceleration. These higher-redshift observations of brighter-than-expected supernovae protect us against any systematic effects that would dim supernovae for reasons other than acceleration.

Prior to the 1998 release of the supernova results there were already several lines of evidence that paved the way for the relatively rapid acceptance of the supernova evidence for the acceleration of the universe. Three in particular included:

In both cases the zero of time corresponds to the present day, and that has been defined so that the slope matches the current expansion rate of the universe (Hubble's constant is taken to be 70 km/s/Mpc). Both types of universe would have initially decelerated, but the universe with the cosmological constant later switched and started accelerating. The cosmological constant universe is older because it took longer to reach its present rate of expansion (13.5 Gyr) than the matter-only universe (9.3 Gyr).

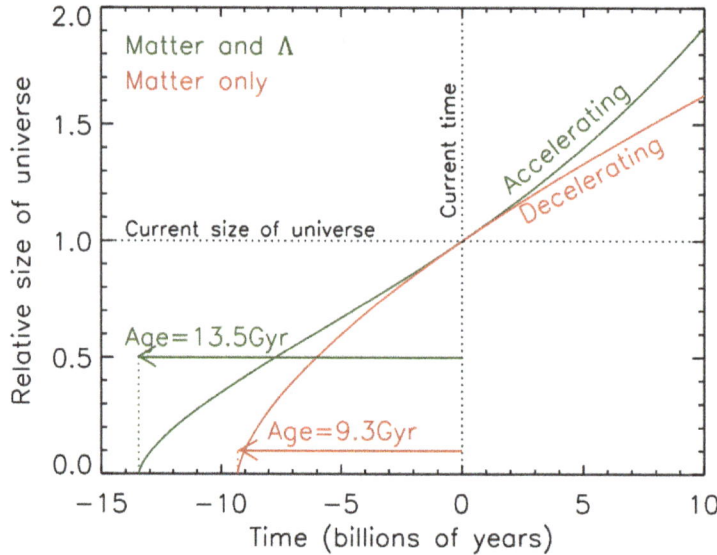

Figure: The relative size of the universe as a function of time for a flat universe made entirely of matter (red) and one made of 30% matter and 70% cosmological constant (green).

- The universe appeared younger than the oldest stars.

  Stellar evolution is well understood, and observations of stars in globular clusters and elsewhere indicate that the oldest stars are over 13 billion years old. We can compare this to the age of the universe by measuring the universe's rate of expansion today and tracing that back to the time of the big bang. If the universe had decelerated to its current speed then the age would be lower than if it had accelerated to its current speed (see Figure 5). A flat universe made only of matter would only be about 9 billion years old -- a major problem given that this is several billion years younger than the oldest stars. On the other hand, a flat universe with 74% cosmological constant would be about 13.7 billion years old. Thus the observation that the universe is currently accelerating solved the age paradox.

- There were too many distant galaxies.

  Galaxy number counts had already been used widely in attempts to estimate the deceleration of the expansion of the universe. The volume of space between two redshifts differs depending on the expansion history of the universe (for a given solid angle). Using the number of galaxies between two redshifts as a measure of the volume of space, observers had measured that distant volumes seemed too large compared with the predictions of a decelerating universe. Either the luminosity of galaxies or the number of galaxies per unit volume was evolving with time in an unexpected way, or the volumes we were calculating were incorrect. An accelerating universe could explain the observations without invoking any strange galaxy evolution.

- The observed flatness of the universe despite insufficient matter.

Using measurements of temperature fluctuations in the cosmic microwave background radiation (CMB) from when the universe was ~380,000 years old one can conclude that the Universe is spatially flat to within a few percent. By combining these data with accurate H_0 measurements and/or measurements of the matter density of the universe, it becomes clear that the matter in the Universe only contributes approximately 23% to the critical density. One way to account for the missing energy density would be to invoke a cosmological constant. As it turns out, the amount of cosmological constant needed to explain the acceleration observed in the supernova data, was just what was needed to also make the universe flat. Therefore the cosmological constant solved the apparent contradiction between the matter-density and CMB observations.

Unresolved Issues

Cosmological Constant Problem

The cosmological constant problem arises because, using naive naturalness arguments in quantum field theory, one cannot explain why the observed cosmological constant is so small. Quantum mechanical calculations that sum the contributions from all vacuum modes below an ultraviolet cutoff at the Planck scale give a vacuum energy density of $\rho_\Lambda \sim 10^{112} erg/cm^3$. This exceeds the cosmologically observed value of $\rho_\Lambda \sim 10^{-8} erg/cm^3$ by about 120 orders of magnitude.

Coincidence Problem

The cosmological constant is not diluted as the universe expands, whereas the density of matter drops in inverse proportion to the volume. This means that there is only a fleeting moment of cosmological time during which the matter density will be of comparable magnitude to the vacuum energy density. Many argue that to be living in that moment is too unlikely to be coincidence. This has been called the coincidence problem, and has motivated theories beyond the cosmological constant with more general forms of dark energy that may change with time.

Dark Energy or Cosmological Constant

These unresolved issues have motivated the current observational effort to test whether the cosmological constant is a valid cause of the acceleration of the universe. Other theories, such as fledgling theories of quantum gravity (e.g. brane-motivated cosmologies), naturally produce dark energy candidates with properties different from the standard cosmological constant (Padmanabhan, 2003). Phenomenological theories such as quintessence have also been proposed, which have a time-varying value of dark energy. Although these models are designed partially to negate the coincidence problem by having dark energy solutions that can track the matter density, these then suffer a new fine-tuning problem as they introduce additional parameters whose values need to be fine-tuned to produce the evolution needed (Weinberg 1989, 2000).

Anthropic Solutions

Many argue that the coincidence problem is most simply solved by anthropic considerations. That is, were the value of the cosmological constant much higher or lower than the observed value it would disrupt structure formation in the universe and humans would not exist. Although many argue against anthropic solutions on philosophical grounds in preference for solutions that invoke some deeper physical principles, anthropic arguments are gaining more prominence, especially in light of the emergence of the string landscape. It had been hoped that string theory would give a well motivated fundamental explanation for the values of the constants of nature, but now it seems ours is only one of many possible solutions, which we find ourselves in by chance constrained by anthropic requirements (Polchinski, 2006).

Dark Gravity

Dark energy also encompasses the possibility that there is no additional energy density component to the universe, but rather that the equations of general relativity need revision. In this sense general relativity might be a limit of a more complete theory of gravity in the same way that Newtonian gravity is a low-energy limit of general relativity. This possibility is also known as dark gravity.

Positive Value

Observations announced in 1998 of distance–redshift relation for Type Ia supernovae indicated that the expansion of the universe is accelerating. When combined with measurements of the cosmic microwave background radiation these implied a value of $\Omega_\Lambda \approx 0.7$, a result which has been supported and refined by more recent measurements. There are other possible causes of an accelerating universe, such as quintessence, but the cosmological constant is in most respects the simplest solution. Thus, the current standard model of cosmology, the Lambda-CDM model, includes the cosmological constant, which is measured to be on the order of 10^{-52} m^{-2}, in metric units. It is often expressed as 10^{-35} s^{-2} or 10^{-122} in other unit systems. The value is based on recent measurements of vacuum energy density, $\rho_{\text{vacuum}} = 5.96 \times 10^{-27}$ kg/m^3,, or 10^{-47} GeV^4, 10^{-29} g/cm^3 in other unit systems.

As was only recently seen, by works of 't Hooft, Susskind and others, a positive cosmological constant has surprising consequences, such as a finite maximum entropy of the observable universe.

Predictions

Quantum Field Theory

A major outstanding problem is that most quantum field theories predict a huge value for the quantum vacuum. A common assumption is that the quantum vacuum is equivalent to the cosmological constant. Although no theory exists that supports this assumption, arguments can be made in its favor.

Such arguments are usually based on dimensional analysis and effective field theory. If the universe is described by an effective local quantum field theory down to the Planck scale, then we would expect a cosmological constant of the order of M_{pl}^2 (6×10^{54} eV^2 in natural unit or 1in reduced Planck unit). As noted above, the measured cosmological constant is smaller than this by a factor of $\sim 10^{-120}$. This discrepancy has been called "the worst theoretical prediction in the history of physics!".

Some supersymmetric theories require a cosmological constant that is exactly zero, which further complicates things. This is the *cosmological constant problem*, the worst problem of fine-tuning in physics: there is no known natural way to derive the tiny cosmological constant used in cosmology from particle physics.

A possible solution is offered by light front quantization, a rigorous alternative to the usual second quantization method. Vacuum fluctuations do not appear in the Light-Front vacuum state. This absence means that there is no contribution from QED, Weak interactions and QCD to the cosmological constant which is thus predicted to be zero in a flat space-time. Unlike supersymmetric theories, that light front quantization predict $\Lambda=0$ within the standard model of particle physics may not be a problem since the small non-zero value of the cosmological constant could originate for example from a slight curvature of the shape of the universe (which is not excluded within 0.4% (as of 2017)) since a curved-space could modify the Higgs field zero-mode, thereby possibly producing a non-zero contribution to the cosmological constant.

Anthropic Principle

One possible explanation for the small but non-zero value was noted by Steven Weinberg in 1987 following the anthropic principle. Weinberg explains that if the vacuum energy took different values in different domains of the universe, then observers would necessarily measure values similar to that which is observed: the formation of life-supporting structures would be suppressed in domains where the vacuum energy is much larger. Specifically, if the vacuum energy is negative and its absolute value is substantially larger than it appears to be in the observed universe (say, a factor of 10 larger), holding all other variables (e.g. matter density) constant, that would mean that the universe is closed; furthermore, its lifetime would be shorter than the age of our universe, possibly too short for intelligent life to form. On the other hand, a universe with a large positive cosmological constant would expand too fast, preventing galaxy formation. According to Weinberg, domains where the vacuum energy is compatible with life would be comparatively rare. Using this argument, Weinberg predicted that the cosmological constant would have a value of less than a hundred times the currently accepted value. In 1992, Weinberg refined this prediction of the cosmological constant to 5 to 10 times the matter density.

This argument depends on a lack of a variation of the distribution (spatial or otherwise) in the vacuum energy density, as would be expected if dark energy were the cosmological constant. There is no evidence that the vacuum energy does vary, but it may be the case if, for example, the vacuum energy is (even in part) the potential of a scalar field such as the residual inflaton. Another theoretical approach that deals with the issue is that of multiverse theories, which predict a large number of "parallel" universes with different laws of physics and/or values of fundamental constants. Again, the anthropic principle states that we can only live in one of the universes that is compatible with some form of intelligent life. Critics claim that these theories, when used as an explanation for fine-tuning, commit the inverse gambler's fallacy.

In 1995, Weinberg's argument was refined by Alexander Vilenkin to predict a value for the cosmological constant that was only ten times the matter density, i.e. about three times the current value since determined.

Equation of State

The relation between the pressure and the energy density (if one exists), $P = P(\rho)$ is called the *equation of state*. For (ideal) matter, radiation and vacuum energy it takes a very simple form,

$$P = w \begin{cases} w = 0 \text{ "}pressureless\text{" }matter\,('dust') \\ w = 1/3 \text{ radiation} \\ w = -1 \text{ vacuum energy or cosmological constant} \end{cases}$$

The Equation

The perfect gas equation of state may be written as

$$p = \rho_m RT = \rho_m C^2$$

where ρ_m is the mass density, R is the particular gas constant, T is the temperature and $C = \sqrt{RT}$ is a characteristic thermal speed of the molecules. Thus

$$w = \frac{p}{\rho} = \frac{\rho_m C^2}{\rho_m c^2} = \frac{C^2}{c^2} \approx 0$$

where c is the speed of light, $\rho = \rho_m c^2$ and $C \ll c$ for a "cold" gas.

FLRW Equations and the Equation of State

The equation of state may be used in Friedmann–Lemaître–Robertson–Walker equations to describe the evolution of an isotropic universe filled with a perfect fluid. If a is the scale factor then

$$\rho \propto a^{-3(1+w)}.$$

If the fluid is the dominant form of matter in a flat universe, then

$$a \propto t^{\frac{2}{3(1+w)}},$$

where t is the proper time.

In general the Friedmann acceleration equation is

$$3\frac{\ddot{a}}{a} = \Lambda - 4\pi G(\rho + 3p)$$

where Λ is the cosmological constant and G is Newton's constant, and $\ddot{a}$ is the second proper time derivative of the scale factor.

If we define (what might be called "effective") energy density and pressure as

$$\rho' \equiv \rho + \frac{\Lambda}{8\pi G}$$

$$p' \equiv p - \frac{\Lambda}{8\pi G}$$

and

$$p' = w' \rho'$$

the acceleration equation may be written as

$$\frac{\ddot{a}}{a} = -\frac{4}{3}\pi G\left(\rho' + 3p'\right) = -\frac{4}{3}\pi G(1 + 3w')\rho'$$

Non-relativistic Matter

The equation of state of ordinary non-relativistic matter (e.g. cold dust) is $w = 0$, which means that it is diluted as $\rho \propto a^{-3} = V^{-1}$, where V is the volume. This means that the energy density red-shifts as the volume, which is natural for ordinary non-relativistic matter.

Ultra-relativistic Matter

The equation of state of ultra-relativistic matter (e.g. radiation, but also matter in the very early universe) is $w = 1/3$ which means that it is diluted as $\rho \propto a^{-4}$. In an expanding universe, the energy density decreases more quickly than the volume expansion, because radiation has momentum and, by the de Broglie hypothesis a wavelength, which is red-shifted.

Acceleration of Cosmic Inflation

Cosmic inflation and the accelerated expansion of the universe can be characterized by the equation of state of dark energy. In the simplest case, the equation of state of the cosmological constant is $w = -1$. In this case, the above expression for the scale factor is not valid and $a \propto e^{Ht}$, where the constant H is the Hubble parameter. More generally, the expansion of the universe is accelerating for any equation of state $w < -1/3$. The accelerated expansion of the Universe was indeed observed. According to observations, the value of equation of state of cosmological constant is near -1.

Hypothetical phantom energy would have an equation of state $w < -1$, and would cause a Big Rip. Using the existing data, it is still impossible to distinguish between phantom $w < -1$ and non-phantom $w \geq -1$.

Fluids

In an expanding universe, fluids with larger equations of state disappear more quickly than those with smaller equations of state. This is the origin of the flatness and monopole problems of the big bang: curvature has $w = -1/3$ and monopoles have $w = 0$, so if they were around at the time of the early big bang, they should still be visible today. These problems are solved by cosmic inflation which has $w \approx -1$. Measuring the equation of state of dark energy is one of the largest efforts of observational cosmology. By accurately measuring w, it is hoped that the cosmological constant could be distinguished from quintessence which has $w \neq -1$.

Scalar modeling

A scalar field ϕ can be viewed as a sort of perfect fluid with equation of state

$$w = \frac{\frac{1}{2}\dot{\phi}^2 - V(\phi)}{\frac{1}{2}\dot{\phi}^2 + V(\phi)}$$

where $\dot{\phi}$ is the time-derivative of ϕ and $V(\phi)$ is the potential energy. A free ($V = 0$) scalar field has $w = 1$, and one with vanishing kinetic energy is equivalent to a cosmological constant: $w = -1$. Any equation of state in between, but not crossing the $w = -1$ barrier known as the Phantom Divide Line (PDL), is achievable, which makes scalar fields useful models for many phenomena in cosmology.

Recombination

As the Universe cools further, a time comes when it is thermodynamically favourable for ions (protons and He^{2+} nuclei) and electrons to combine and form neutral atoms. This is the epoch of recombination, the next important transition in the history of our Universe. With the rapidly diminishing density of free electrons, the photon scattering rate, $\Gamma_{T,e}$ drops below the expansion rate H, the photons decouple from the electrons and can stream freely (their mean free path becomes very much longer): the Universe is now transparent to radiation. Thus, as we look back in time with even our most powerful photon-collecting telescopes, the epoch of recombination is the ultimate frontier, the furthest location and the earliest time we can reach with electromagnetic radiation. Once photon and baryons have decoupled, the latter are no longer compelled to have the the same temperature as the photons.

The temperature at which recombination takes place depends on the baryon-to-photon ratio, η, _ and on the ionisation potential of the species involved. For simplicity, we shall limit ourselves to H with ionisation potential $Q = 13.6\,eV$ from the ground state, and ignore He with ionisation potentials of 24.6 eV and 54.4 eV to form He^+ and He^{2+} respectively.

Before recombination, the reaction in question:

$$H + \gamma \rightleftharpoons p + e^-$$

is in statistical equilibrium, with the photoionization rate balancing the radiative recombination rate. In statistical equilibrium at temperature T, the number density n_x of particles with mass m_x is given by the Maxwell-Boltzmann equation:

$$n_x = g_x \left(\frac{m_x \kappa T}{2\pi\hbar^2} \right)^{3/2} \exp\left[-\frac{m_x c^2}{\kappa T} \right]$$

where g_x is the statistical weight of particle X This expression applies in the non-relativistic regime, i.e. when $\kappa T \ll m_x c^2$.

Writing equation above for H atoms, protons and free electrons, we can construct an equation that relates the number densities of these particles:

$$\frac{n_H}{n_p n_e} = \frac{g_H}{g_p g_e} \left(\frac{m_H}{m_p m_e} \right)^{3/2} \left(\frac{\kappa T}{2\pi\hbar^2} \right)^{-3/2} \exp\left[\frac{\left(m_p + m_e - m_H \right) c^2}{\kappa T} \right]$$

Equation above can be simplified further considering that: (i) the ratio of the statistical weights is 1; (ii) $m_H \simeq m_p$ and (iii) the term in the numerator of the exponential factor is the binding energy of the H atom, i.e. the ionisation potential Q. With these simplifications, we obtain the Saha equation:

$$\frac{n_H}{n_p n_e} = \left(\frac{m_e \kappa T}{2\pi\hbar^2} \right)^{-3/2} \exp\left[\frac{Q}{\kappa T} \right]$$

What we want to do now is to use the Saha equation to deduce the ionisation fraction:

$$X \equiv \frac{n_p}{n_p + n_H} = \frac{n_p}{n_b} = \frac{n_e}{n_b}$$

as a function of T and n With the above definition (which implicitly assumes charge neutrality in the Universe), $X = 1$ when the baryons are fully ionised $X = 0.5$ when half of the baryons are ionised, and X = 0 when the baryons are all in neutral atoms.

With the substitutions:

$$n_H = \frac{1-X}{X} n_p, \qquad n_e = n_p,$$

we can re-write equation above as:

$$\frac{1-X}{X} = n_p \left(\frac{m_e \kappa T}{2\pi\hbar^2} \right)^{-3/2} \exp\left[\frac{Q}{\kappa T} \right]$$

We can express n_p in terms of η

$$\eta = \frac{n_p}{X n_\gamma}$$

where for a blackbody spectrum:

$$n_\gamma = \frac{2.404}{\pi^2}\left(\frac{\kappa T}{\hbar c}\right)^3 = 0.244\left(\frac{\kappa T}{\hbar c}\right)^3$$

so that:

$$n_p = 0.244 X \eta \left(\frac{\kappa T}{\hbar c}\right)^3$$

which we can now substitute into re-written equation to give

$$\frac{1-X}{X} = 3.84\eta \left(\frac{\kappa T}{m_e c^2}\right)^{3/2} \exp\left[\frac{Q}{\kappa T}\right]$$

Solving the quadratic equation in X, we find that the positive root is

$$X = \frac{-1+\sqrt{1+4S}}{2S}$$

Where

$$S(T,\eta) = 3.84\eta \left(\frac{\kappa T}{m_e c^2}\right)^{3/2} \exp\left[\frac{Q}{\kappa T}\right]$$

Note that when $\gg Q, X \simeq 1$ and the gas is close to being fully ionised. Once κT falls below $Q, X \to 0$; however, both _ and the term $\left(\kappa T/m_e c^2\right)^{3/2}$ are small numbers, and their product is overcome by the exponential term only once the temperature has fallen well below the binding energy.

We can solve above equation numerically to find the value of T when $X = 0.5$ which we define to be the epoch of recombination (half of the baryons ionised and half of them neutral). With $\eta = 6.1\times10^{-10}$, we have:

$$\kappa T_{rec} = 0.323\ eV = \frac{Q}{42}$$

Scaling back $T_{CMB,0} = 2.7255\,K$ to $T_{rec} = 0.323\ eV \equiv 3750\,K$, we and $(1+z_{rec}) = 1375$,

which corresponds to time $t_{rec} = 251\,000\ yr$

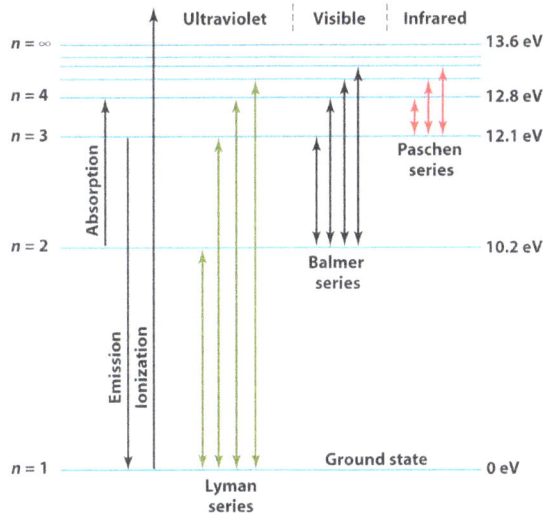

Figure: Energy levels of the H atom.

References

- Partridge, R. B. (2007). 3K: The Cosmic Microwave Background Radiation (illustrated ed.). Cambridge University Press. p. xvii. ISBN 978-0-521-35808-8

- Peebles, P. J. E.; Ratra, Bharat (2003). "The cosmological constant and dark energy". Reviews of Modern Physics. 75 (2): 559–606. arXiv:astro-ph/0207347. Bibcode:2003RvMP...75..559P. doi:10.1103/RevModPhys.75.559

- Clavin, Whitney; Jenkins, Ann; Villard, Ray (7 January 2014). "NASA's Hubble and Spitzer Team up to Probe Faraway Galaxies". NASA. Retrieved 8 January 2014

- Yao, W. M.; et al. (2006). "Review of Particle Physics: Dark Matter" (PDF). Journal of Physics G. 33 (1): 1–1232. arXiv:astro-ph/0601168. Bibcode:2006JPhG...33....1Y. doi:10.1088/0954-3899/33/1/001

- Weinberg, Steven (1993). Dreams of a Final Theory: the search for the fundamental laws of nature. Vintage Press. p. 182. ISBN 0-09-922391-0

- Rosen, Rebecca J. "Einstein Likely Never Said One of His Most Oft-Quoted Phrases". The Atlantic. The Atlantic Media Company. Retrieved 10 August 2013

- H. Leutwyler, J.R. Klauder, L. Streit. Quantum field theory on lightlike slabs, Nuovo Cim. A66 (1970) 536 DOI: 10.1007/BF02826338

Chapter 3

Physical Cosmology

The study of the large-scale structures of the universe, and their dynamics is under the scope of physical cosmology. This chapter has been carefully written to provide an easy understanding of physical cosmology through the principal topics like cosmological principle, Hubble's law, baryogenesis and cosmological perturbation theory.

The universe has captured the human imagination since the ancient ages. Its study emerged with the association of the motion of celestial objects in the sky with divinity. Many early attempts to track the motion of the sun and the moon are recorded in history. The Nebra sky disc dating back to the Bronze Age which marked one of the earliest attempts by humankind to catalogue time by devising a lunisolar calendar, understanding of the periodic nature of celestial events by the ancient Babylonians or the calculations of the periods of planets and eclipses by the Indians of the Indus Valley signify scientific inquiry at its best. Such evidences of early attempts to know the universe are being uncovered still.

Physical cosmology as we know of today, has moved beyond the study of the motions of planets and the sun, to embrace the study of the large-scale dynamics and structures of the universe. It strives to answer the more fundamental questions pertaining to the beginning of the universe, its evolutionary path and the future or possible end. The more definitive origin of modern cosmology begins with all probability with the formulation of the Einstein's general theory of relativity, which gave the theoretical basis for the treatment of space-time in the presence of matter. It laid down the foundation of the most significant work pertaining to the early universe, the theory of expansion and the Big Bang theory, which theorizes that the universe expanded from a singularity around 13.8 billion years ago. A modern dimension to this theory is the hypothesis that the universe underwent a period of accelerated expansion, right after the Big Bang, a model that is called the cosmic inflation theory. A remarkable discovery of our times in this field is the discovery of gravitational waves in 2016, by the LIGO and Virgo Scientific Collaboration, which can provide valuable insight into the physics of the early universe.

Inspite of the great strides made in physical cosmology in the past years, there still remains many perplexing questions to which physics still can provide no definitive answers. The understanding of the big bang singularity or the trigger for the event is too vague and the alternatives to the big bang model are still being explored. A most challenging problem in cosmology and particle physics is the nature of dark energy and dark matter. The eventual fate of the universe remains a matter of conjecture, though theories like the Big Rip or the Big Freeze have been suggested. But these remain weak shots in the dark, and the truth remains that even today, the universe is largely unknown.

Energy of the Cosmos

The lightest chemical elements, primarily hydrogen and helium, were created during the Big Bang through the process of nucleosynthesis. In a sequence of stellar nucleosynthesis reactions, smaller atomic nuclei are then combined into larger atomic nuclei, ultimately forming stable iron group elements such as iron and nickel, which have the highest nuclear binding energies. The net process results in a *later energy release*, meaning subsequent to the Big Bang. Such reactions of nuclear particles can lead to *sudden energy releases* from cataclysmic variables such as novae. Gravitational collapse of matter into black holes also powers the most energetic processes, generally seen in the nuclear regions of galaxies, forming *quasars* and *active galaxies*.

Cosmologists cannot explain all cosmic phenomena exactly, such as those related to the accelerating expansion of the universe, using conventional forms of energy. Instead, cosmologists propose a new form of energy called dark energy that permeates all space. One hypothesis is that dark energy is just the vacuum energy, a component of empty space that is associated with the virtual particles that exist due to the uncertainty principle.

There is no clear way to define the total energy in the universe using the most widely accepted theory of gravity, general relativity. Therefore, it remains controversial whether the total energy is conserved in an expanding universe. For instance, each photon that travels through intergalactic space loses energy due to the redshift effect. This energy is not obviously transferred to any other system, so seems to be permanently lost. On the other hand, some cosmologists insist that energy is conserved in some sense; this follows the law of conservation of energy.

Thermodynamics of the universe is a field of study that explores which form of energy dominates the cosmos – relativistic particles which are referred to as radiation, or non-relativistic particles referred to as matter. Relativistic particles are particles whose rest mass is zero or negligible compared to their kinetic energy, and so move at the speed of light or very close to it; non-relativistic particles have much higher rest mass than their energy and so move much slower than the speed of light.

As the universe expands, both matter and radiation in it become diluted. However, the energy densities of radiation and matter dilute at different rates. As a particular volume expands, mass energy density is changed only by the increase in volume, but the energy density of radiation is changed both by the increase in volume and by the increase in the wavelength of the photons that make it up. Thus the energy of radiation becomes a smaller part of the universe's total energy than that of matter as it expands. The very early universe is said to have been 'radiation dominated' and radiation controlled the deceleration of expansion. Later, as the average energy per photon becomes roughly 10 eV and lower, matter dictates the rate of deceleration and the universe is said to be 'matter dominated'. The intermediate case is not treated well analytically. As the expansion of the universe continues, matter dilutes even further and the cosmological constant becomes dominant, leading to an acceleration in the universe's expansion.

Areas of Study

Below, some of the most active areas of inquiry in cosmology are described, in roughly chronological order. This does not include all of the Big Bang cosmology, which is presented in *Timeline of the Big Bang*.

Very Early Universe

The early, hot universe appears to be well explained by the Big Bang from roughly 10^{-33} seconds onwards, but there are several problems. One is that there is no compelling reason, using current particle physics, for the universe to be flat, homogeneous, and isotropic. Moreover, grand unified theories of particle physics suggest that there should be magnetic monopoles in the universe, which have not been found. These problems are resolved by a brief period of cosmic inflation, which drives the universe to flatness, smooths out anisotropies and inhomogeneities to the observed level, and exponentially dilutes the monopoles. The physical model behind cosmic inflation is extremely simple, but it has not yet been confirmed by particle physics, and there are difficult problems reconciling inflation and quantum field theory. Some cosmologists think that string theory and brane cosmology will provide an alternative to inflation.

Another major problem in cosmology is what caused the universe to contain far more matter than antimatter. Cosmologists can observationally deduce that the universe is not split into regions of matter and antimatter. If it were, there would be X-rays and gamma rays produced as a result of annihilation, but this is not observed. Therefore, some process in the early universe must have created a small excess of matter over antimatter, and this (currently not understood) process is called *baryogenesis*. Three required conditions for baryogenesis were derived by Andrei Sakharov in 1967, and requires a violation of the particle physics symmetry, called CP-symmetry, between matter and antimatter. However, particle accelerators measure too small a violation of CP-symmetry to account for the baryon asymmetry. Cosmologists and particle physicists look for additional violations of the CP-symmetry in the early universe that might account for the baryon asymmetry.

Both the problems of baryogenesis and cosmic inflation are very closely related to particle physics, and their resolution might come from high energy theory and experiment, rather than through observations of the universe.

Big Bang Theory

Big Bang nucleosynthesis is the theory of the formation of the elements in the early universe. It finished when the universe was about three minutes old and its temperature dropped below that at which nuclear fusion could occur. Big Bang nucleosynthesis had a brief period during which it could operate, so only the very lightest elements were produced. Starting from hydrogen ions (protons), it principally produced deuterium, helium-4, and lithium. Other elements were produced in only trace abundances. The basic theory of nucleosynthesis was developed in 1948 by George Gamow, Ralph Asher Alpher, and Robert Herman. It was used for many years as a probe of physics at the time of the Big Bang, as the theory of Big Bang nucleosynthesis connects the abundances of primordial light elements with the features of the early universe. Specifically, it can be used to test the equivalence principle, to probe dark matter, and test neutrino physics. Some cosmologists have proposed that Big Bang nucleosynthesis suggests there is a fourth "sterile" species of neutrino.

Standard Model of Big Bang Cosmology

The ΛCDM (Lambda cold dark matter) or Lambda-CDM model is a parametrization of the Big Bang cosmological model in which the universe contains a cosmological constant, denoted by

Lambda (Greek Λ), associated with dark energy, and cold dark matter (abbreviated CDM). It is frequently referred to as the standard model of Big Bang cosmology.

Cosmic Microwave Background

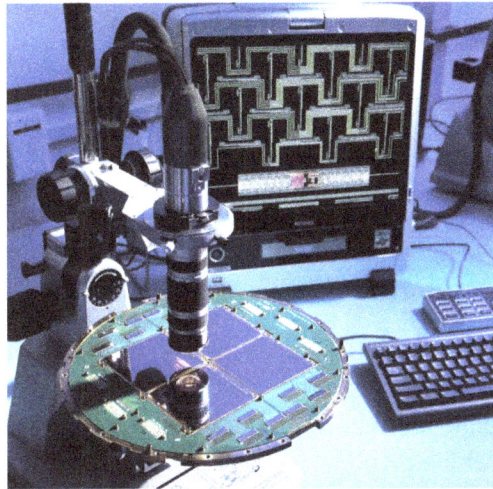

Evidence of gravitational waves in the infant universe may have been uncovered by the microscopic examination of the focal plane of the BICEP2 radio telescope.

The cosmic microwave background is radiation left over from decoupling after the epoch of recombination when neutral atoms first formed. At this point, radiation produced in the Big Bang stopped Thomson scattering from charged ions. The radiation, first observed in 1965 by Arno Penzias and Robert Woodrow Wilson, has a perfect thermal black-body spectrum. It has a temperature of 2.7 kelvins today and is isotropic to one part in 10^5. Cosmological perturbation theory, which describes the evolution of slight inhomogeneities in the early universe, has allowed cosmologists to precisely calculate the angular power spectrum of the radiation, and it has been measured by the recent satellite experiments (COBE and WMAP) and many ground and balloon-based experiments (such as Degree Angular Scale Interferometer, Cosmic Background Imager, and Boomerang). One of the goals of these efforts is to measure the basic parameters of the Lambda-CDM model with increasing accuracy, as well as to test the predictions of the Big Bang model and look for new physics. The results of measurements made by WMAP, for example, have placed limits on the neutrino masses.

Newer experiments, such as QUIET and the Atacama Cosmology Telescope, are trying to measure the polarization of the cosmic microwave background. These measurements are expected to provide further confirmation of the theory as well as information about cosmic inflation, and the so-called secondary anisotropies, such as the Sunyaev-Zel'dovich effect and Sachs-Wolfe effect, which are caused by interaction between galaxies and clusters with the cosmic microwave background.

On 17 March 2014, astronomers of the BICEP2 Collaboration announced the apparent detection of *B*-mode polarization of the CMB, considered to be evidence of primordial gravitational waves that are predicted by the theory of inflation to occur during the earliest phase of the Big Bang. However, later that year the Planck collaboration provided a more accurate measurement of cosmic dust, concluding that the B-mode signal from dust is the same strength as that reported from BICEP2.

On January 30, 2015, a joint analysis of BICEP2 and Planck data was published and the European Space Agency announced that the signal can be entirely attributed to interstellar dust in the Milky Way.

Formation and Evolution of Large-scale Structure

Understanding the formation and evolution of the largest and earliest structures (i.e., quasars, galaxies, clusters and superclusters) is one of the largest efforts in cosmology. Cosmologists study a model of hierarchical structure formation in which structures form from the bottom up, with smaller objects forming first, while the largest objects, such as superclusters, are still assembling. One way to study structure in the universe is to survey the visible galaxies, in order to construct a three-dimensional picture of the galaxies in the universe and measure the matter power spectrum. This is the approach of the *Sloan Digital Sky Survey* and the 2dF Galaxy Redshift Survey.

Another tool for understanding structure formation is simulations, which cosmologists use to study the gravitational aggregation of matter in the universe, as it clusters into filaments, super-clusters and voids. Most simulations contain only non-baryonic cold dark matter, which should suffice to understand the universe on the largest scales, as there is much more dark matter in the universe than visible, baryonic matter. More advanced simulations are starting to include baryons and study the formation of individual galaxies. Cosmologists study these simulations to see if they agree with the galaxy surveys, and to understand any discrepancy.

Other, complementary observations to measure the distribution of matter in the distant universe and to probe reionization include:

* The Lyman-alpha forest, which allows cosmologists to measure the distribution of neutral atomic hydrogen gas in the early universe, by measuring the absorption of light from distant quasars by the gas.

* The 21 centimeter absorption line of neutral atomic hydrogen also provides a sensitive test of cosmology.

* Weak lensing, the distortion of a distant image by gravitational lensing due to dark matter.

These will help cosmologists settle the question of when and how structure formed in the universe.

Dark Matter

Evidence from Big Bang nucleosynthesis, the cosmic microwave background, structure formation, and galaxy rotation curves suggests that about 23% of the mass of the universe consists of non-baryonic dark matter, whereas only 4% consists of visible, baryonic matter. The gravitational effects of dark matter are well understood, as it behaves like a cold, non-radiative fluid that forms haloes around galaxies. Dark matter has never been detected in the laboratory, and the particle physics nature of dark matter remains completely unknown. Without observational constraints, there are a number of candidates, such as a stable supersymmetric particle, a weakly interacting massive particle, a gravitationally-interacting massive particle, an axion, and a massive compact halo object. Alternatives to the dark matter hypothesis include a modification of gravity at small accelerations (MOND) or an effect from brane cosmology.

Dark Energy

If the universe is flat, there must be an additional component making up 73% (in addition to the 23% dark matter and 4% baryons) of the energy density of the universe. This is called dark energy. In order not to interfere with Big Bang nucleosynthesis and the cosmic microwave background, it must not cluster in haloes like baryons and dark matter. There is strong observational evidence for dark energy, as the total energy density of the universe is known through constraints on the flatness of the universe, but the amount of clustering matter is tightly measured, and is much less than this. The case for dark energy was strengthened in 1999, when measurements demonstrated that the expansion of the universe has begun to gradually accelerate.

Apart from its density and its clustering properties, nothing is known about dark energy. *Quantum field theory* predicts a cosmological constant (CC) much like dark energy, but 120 orders of magnitude larger than that observed. Steven Weinberg and a number of string theorists have invoked the 'weak anthropic principle': i.e. the reason that physicists observe a universe with such a small cosmological constant is that no physicists (or any life) could exist in a universe with a larger cosmological constant. Many cosmologists find this an unsatisfying explanation: perhaps because while the weak anthropic principle is self-evident (given that living observers exist, there must be at least one universe with a cosmological constant which allows for life to exist) it does not attempt to explain the context of that universe. For example, the weak anthropic principle alone does not distinguish between:

- Only one universe will ever exist and there is some underlying principle that constrains the CC to the value we observe.

- Only one universe will ever exist and although there is no underlying principle fixing the CC, we got lucky.

- Lots of universes exist (simultaneously or serially) with a range of CC values, and of course ours is one of the life-supporting ones.

Other possible explanations for dark energy include quintessence or a modification of gravity on the largest scales. The effect on cosmology of the dark energy that these models describe is given by the dark energy's equation of state, which varies depending upon the theory. The nature of dark energy is one of the most challenging problems in cosmology.

A better understanding of dark energy is likely to solve the problem of the ultimate fate of the universe. In the current cosmological epoch, the accelerated expansion due to dark energy is preventing structures larger than superclusters from forming. It is not known whether the acceleration will continue indefinitely, perhaps even increasing until a big rip, or whether it will eventually reverse, lead to a big freeze, or follow some other scenario.

Gravitational Waves

Gravitational waves are ripples in the curvature of spacetime that propagate as waves at the speed of light, generated in certain gravitational interactions that propagate outward from their source. Gravitational-wave astronomy is an emerging branch of observational astronomy which aims to use gravitational waves to collect observational data about sources of detectable gravitational waves such as binary star systems composed of white dwarfs, neutron stars, and black holes; and events such as supernovae, and the formation of the early universe shortly after the Big Bang.

In 2016, the LIGO Scientific Collaboration and Virgo Collaboration teams announced that they had made the first observation of gravitational waves, originating from a pair of merging black holes using the Advanced LIGO detectors. On June 15, 2016, a second detection of gravitational waves from coalescing black holes was announced. Besides LIGO, many other gravitational-wave observatories (detectors) are under construction.

Other Areas of Inquiry

Cosmologists also study:

- Whether primordial black holes were formed in our universe, and what happened to them.

- Detection of cosmic rays with energies above the GZK cutoff, and whether it signals a failure of special relativity at high energies.

- The equivalence principle, whether or not Einstein's general theory of relativity is the correct theory of gravitation, and if the fundamental laws of physics are the same everywhere in the universe.

- The increasing complexity of universal structures, an example being the progressively greater energy rate density.

Cosmological Principle

The Cosmology Principle, introduced by Einstein, states that the Universe is homogeneous and isotropic. This only applies to 'sufficiently' large scales: on smaller scales, matter is seen to cluster in galaxies, which themselves cluster in groups, clusters and super-clusters. We think that these structures grew gravitationally from very small `primordial perturbations In the inflationary paradigm, these primordial perturbations were at one stage quantum fluctuations, which were inflated to macroscopic scales. When gravity acted on these small perturbations, it made denser regions even more dense, and under dense regions even more under dense, resulting in the structures.

Homogeneous means uniform or evenly distributed. Clearly, this is not strictly true, since galaxies are clustered. If gravity had not formed clumps on the scales of planets and stars and galaxies, we would not be here! In cosmology, however, homogenous does not mean that all regions of space should appear identical or be smoothly filled with particles. It only means that the same types of structures — stars, galaxies, clusters, and superclusters — are seen everywhere.

The universe does appear smooth or homogeneous on scales larger than about 300 Mpc. Viewed up close, a beach consists of grains of sand and shells and pebbles of many different sizes. From afar, all we see is a beach. The universe is isotropic if it looks the same in all directions. In other words, no observation can be made that will identify an edge or a center. The concept of isotropy is supported by the fact that galaxies do not bunch up in any direction in the sky and by the fact that we observe the same Hubble relation in different directions in the sky. Large telescopes have been used to count faint and distant galaxies in different directions and the numbers are always statistically the same.

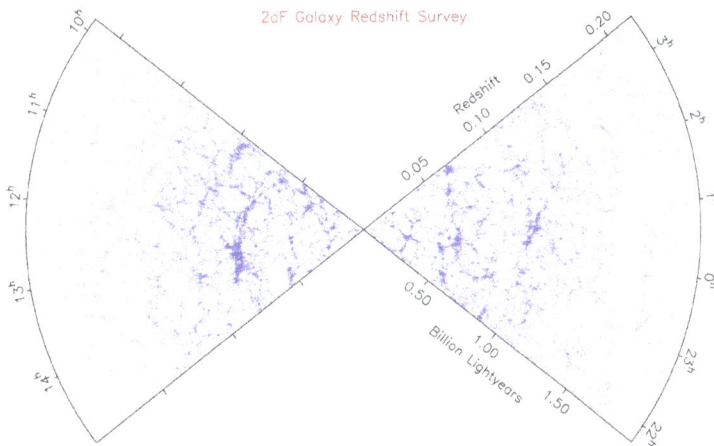

Location of galaxies as a function of redshift from the 2df galaxy survey

It is difficult to test the cosmological principle. The isotropy of the universe is reasonably well confirmed, because observers looking in different directions from the Earth see essentially the same motions and structures. However, we cannot test homogeneity because we cannot travel to distant locations to see if things look any different. When we do look to great distances, we are also looking back to a time when the universe was hotter and denser, so the situations may not be comparable. We cannot compare like with like; nearby galaxies are seen as they are but distant galaxies are seen as they were. All available evidence supports this principle, but our certainty is not very high.

Definition

Astronomer William Keel explains:

> The cosmological principle is usually stated formally as 'Viewed on a sufficiently large scale, the properties of the universe are the same for all observers.' This amounts to the strongly philosophical statement that the part of the universe which we can see is a fair sample, and that the same physical laws apply throughout. In essence, this in a sense says that the universe is knowable and is playing fair with scientists.

The cosmological principle depends on a definition of "observer," and contains an implicit qualification and two testable consequences.

"Observers" means any observer at any location in the universe, not simply any human observer at any location on Earth: as Andrew Liddle puts it, "the cosmological principle [means that] the universe looks the same whoever and wherever you are."

The qualification is that variation in physical structures can be overlooked, provided this does not imperil the uniformity of conclusions drawn from observation: the Sun is different from the Earth, our galaxy is different from a black hole, some galaxies advance toward rather than recede from us, and the universe has a "foamy" texture of galaxy clusters and voids, but none of these different structures appears to violate the basic laws of physics.

The two testable structural consequences of the cosmological principle are homogeneity and isotropy. Homogeneity means that the same observational evidence is available to observers at different locations in the universe ("the part of the universe which we can see is a fair sample"). Isotropy means that the same observational evidence is available by looking in any direction in the universe ("the same physical laws apply throughout"). The principles are distinct but closely related, because a universe that appears isotropic from any two (for a spherical geometry, three) locations must also be homogeneous.

Implications

Observations show that more distant galaxies are closer together and have lower content of chemical elements heavier than lithium. Applying the cosmological principle, this suggests that heavier elements were not created in the Big Bang but were produced by nucleosynthesis in giant stars and expelled across a series of supernovae explosions and new star formation from the supernovae remnants, which means heavier elements would accumulate over time. Another observation is that the furthest galaxies (earlier time) are often more fragmentary, interacting and unusually shaped than local galaxies (recent time), suggesting evolution in galaxy structure as well.

A related implication of the cosmological principle is that the largest discrete structures in the universe are in mechanical equilibrium. Homogeneity and isotropy of matter at the largest scales would suggest that the largest discrete structures are parts of a single indiscrete form, like the crumbs which make up the interior of a cake. At extreme cosmological distances, the property of mechanical equilibrium in surfaces lateral to the line of sight can be empirically tested; however, under the assumption of the cosmological principle, it cannot be detected parallel to the line of sight.

Cosmologists agree that in accordance with observations of distant galaxies, a universe must be non-static if it follows the cosmological principle. In 1923, Alexander Friedmann set out a variant of Einstein's equations of general relativity that describe the dynamics of a homogeneous isotropic universe. Independently, Georges Lemaître derived in 1927 the equations of an expanding universe from the General Relativity equations. Thus, a non-static universe is also implied, independent of observations of distant galaxies, as the result of applying the cosmological principle to general relativity.

Observations

Although the universe is inhomogeneous at smaller scales, it *is* statistically homogeneous on scales larger than 250 million light years. The cosmic microwave background is isotropic, that is to say that its intensity is about the same whichever direction we look at.

However, recent findings have called this view into question. Data from the Planck Mission shows hemispheric bias in 2 respects: one with respect to average temperature (i.e. temperature fluctuations), the second with respect to larger variations in the degree of perturbations (i.e. densities). Therefore, the European Space Agency (the governing body of the Planck Mission) has concluded that these anisotropies are, in fact, statistically significant and can no longer be ignored.

Inconsistencies

The *cosmological principle* implies that at a sufficiently large scale, the universe is homogeneous. This means that different places will appear similar to one another, so sufficiently large structures cannot exist. Yadav and his colleagues have suggested a maximum scale of 260/h Mpc for structures within the universe according to this heuristic. Other authors have suggested values as low as 60/h Mpc. Yadav's calculation suggests that the maximum size of a structure can be about 370 Mpc.

A number of observations conflict with predictions of maximal structure sizes:

- The Clowes–Campusano LQG, discovered in 1991, has a length of 580 Mpc, and is marginally larger than the consistent scale.

- The Sloan Great Wall, discovered in 2003, has a length of 423 Mpc, which is only just consistent with the cosmological principle.

- U1.11, a large quasar group discovered in 2011, has a length of 780 Mpc, and is two times larger than the upper limit of the homogeneity scale.

- The Huge-LQG, discovered in 2012, is three times longer than, and twice as wide as is predicted possible according to these current models, and so challenges our understanding of the universe on large scales.

- In November 2013, a new structure 10 billion light years away measuring 2000-3000 Mpc (more than seven times that of the SGW) has been discovered, the Hercules–Corona Borealis Great Wall, putting further doubt on the validity of the cosmological principle.

In September 2016, however, studies of the expansion of the Universe that have used data taken by the *Planck* mission show it to be highly isotropical, reinforcing the cosmological principle.

Perfect Cosmological Principle

The perfect cosmological principle is an extension of the cosmological principle, and states that the universe is homogeneous and isotropic in space *and* time. In this view the universe looks the same everywhere (on the large scale), the same as it always has and always will. The perfect cosmological principle underpins Steady State theory and emerges from chaotic inflation theory.

Hubble's Law

Hubble's law is a statement of a direct correlation between the distance to a galaxy and its recessional velocity as determined by the red shift. It can be stated as

$$v = H_0 r$$

v = recessional velocity

H_0 = Hubble

r = distance

The reported value of the Hubble parameter has varied widely over the years, testament to the difficulty of astronomical distance measurement. But with high precision experiments after 1990 the range of the reported values has narrowed greatly to values in the range

$$H_0 = \frac{70km/s}{Mpc} \approx \frac{70km/s}{Mly} \quad Mpc = \text{million parsecs}$$

$$Mly = \text{million light years}$$

An often mentioned problem for the Hubble law is Stefan's Quintet. Four of these five stars have similar red shifts but the fifth is quite different, and they appear to be interacting.

The Particle Data Group documents quote a "best modern value" of the Hubble parameter as 72 km/s per megaparsec (+/- 10%). This value comes from the use of type Ia supernovae (which give relative distances to about 5%) along with data from Cepheid variables gathered by the Hubble Space Telescope. The WMAP mission data leads to a Hubble constant of 71 +/- 5% km/s per megaparsec.

Hubble Parameter

The proportionality between recession velocity and distance in the Hubble Law is called the Hubble constant, or more appropriately the Hubble parameter we have a history of revising it. In recent years the value of the Hubble parameter has been considerably refined, and the current value given by the WMAP mission is 71 km/s per megaparsec.

The recession velocities of distant galaxies are known from the red shift, but the distances are much more uncertain. Distance measurement to nearby galaxies uses Cepheid variables as the main standard candle, but more distant galaxies must be examined to determine the Hubble constant since the direct Cepheid distances are all within the range of the gravitational pull of the local cluster. Use of the Hubble Space Telescope has permitted the detection of Cepheid variables in the Virgo cluster which have contributed to refinement of the distance scale.

The Particle Data Group documents quote a "best modern value" of the Hubble constant as 72 km/s per megaparsec (+/- 10%). This value comes from the use of type Ia supernovae (which give relative distances to about 5%) along with data from Cepheid variables gathered by the Hubble Space Telescope. The value from the WMAP survey is 71 km/s per megaparsec.

Another approach to the Hubble parameter gives emphasis to the fact that space itself is expanding, and at any given time can be described by a dimensionless scale factor R(t). The Hubble parameter is the ratio of the rate of change of the scale factor to the current value of the scale factor R:

$$R(t) = \text{dimensionless scale}$$
$$\text{factor for the expanding}$$
$$H(t) = \frac{\dfrac{dR(t)}{dt}}{R(t)} \quad \text{universe}$$
$$R(t_0) = 1$$
$$\text{The scalefactor is} = 1$$
$$\text{at the present time.}$$

The scale factor R for a given observed object in the expanding universe relative to $R_0 = 1$ at the present time may be implied from the z parameter expression of the redshift. The Hubble parameter has the dimensions of inverse time, so a Hubble time t_H may be obtained by inverting the present value of the Hubble parameter.

$$H_0 = 71\frac{km/s}{mpc} = 2.3x10^{-18}\ s^{-1}$$

$$t_H = \frac{1}{2.3x10^{-18}\ s^{-1}} = 13.8x10^9\ years$$

One must use caution in interpreting this "Hubble time" since the relationship of the expansion time to the Hubble time is different for the radiation dominated era and the mass dominated era. Projections of the expansion time may be made from the expansion models.

Hubble Parameter and Red Shifts

The Hubble Law states that the distance to a given galaxy is proportional to the recessional velocity as measured by the Doppler red shift. The red shift of the spectral lines is commonly expressed in terms of the z-parameter, which is the fractional shift in the spectral wavelength. The Hubble distance is given by

$$r = \frac{v}{H_0} = \frac{\beta c}{H_0} = \left[\frac{(z+1)^2-1}{(z+1)^2+1}\right]\frac{c}{H_0}$$

Discovery

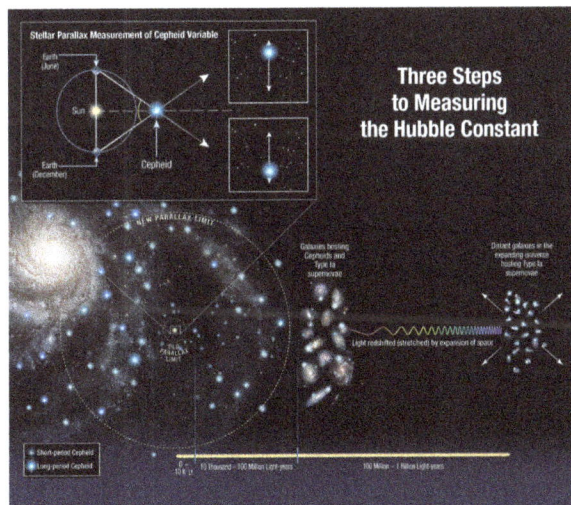

Three steps to the Hubble constant.

A decade before Hubble made his observations, a number of physicists and mathematicians had established a consistent theory of the relationship between space and time by using Einstein's field equations of general relativity. Applying the most general principles to the nature of the universe yielded a dynamic solution that conflicted with the then-prevailing notion of a static universe.

FLRW Equations

In 1922, Alexander Friedmann derived his Friedmann equations from Einstein's field equations, showing that the Universe might expand at a rate calculable by the equations. The parameter used by Friedmann is known today as the scale factor which can be considered as a scale invariant form of the proportionality constant of Hubble's law. Georges Lemaître independently found a similar solution in 1927. The Friedmann equations are derived by inserting the metric for a homogeneous and isotropic universe into Einstein's field equations for a fluid with a given density and pressure. This idea of an expanding spacetime would eventually lead to the Big Bang and Steady State theories of cosmology.

Lemaître's Equation

In 1927, two years before Hubble published his own article, the Belgian priest and astronomer Georges Lemaître was the first to publish research deriving what is now known as Hubble's Law. According to the Canadian astronomer Sidney van den Bergh, "The 1927 discovery of the expansion of the Universe by Lemaître was published in French in a low-impact journal. In the 1931 high-impact English translation of this article a critical equation was changed by omitting reference to what is now known as the Hubble constant.". It is now known that the alterations in the translated paper were carried out by Lemaitre himself.

Shape of the Universe

Before the advent of modern cosmology, there was considerable talk about the size and shape of the universe. In 1920, the Shapley-Curtis debate took place between Harlow Shapley and Heber D. Curtis over this issue. Shapley argued for a small universe the size of the Milky Way galaxy and Curtis argued that the Universe was much larger. The issue was resolved in the coming decade with Hubble's improved observations.

Cepheid Variable Stars Outside of the Milky Way

Edwin Hubble did most of his professional astronomical observing work at Mount Wilson Observatory, home to the world's most powerful telescope at the time. His observations of Cepheid variable stars in spiral nebulae enabled him to calculate the distances to these objects. Surprisingly, these objects were discovered to be at distances which placed them well outside the Milky Way. They continued to be called "nebulae" and it was only gradually that the term "galaxies" took over.

Combining Redshifts with Distance Measurements

The parameters that appear in Hubble's law, velocities and distances, are not directly measured. In reality we determine, say, a supernova brightness, which provides information about its distance, and the redshift $z = \Delta\lambda/\lambda$ of its spectrum of radiation. Hubble correlated brightness and parameter z.

Combining his measurements of galaxy distances with Vesto Slipher and Milton Humason's measurements of the redshifts associated with the galaxies, Hubble discovered a rough proportionality between redshift of an object and its distance. Though there was considerable scatter (now known to be caused by peculiar velocities – the 'Hubble flow' is used to refer to the region of space far enough out that the recession velocity is larger than local peculiar velocities), Hubble was able to plot a trend line from the 46 galaxies he studied and obtain a value for the Hubble constant of 500 km/s/Mpc (much higher than the currently accepted value due to errors in his distance calibrations).

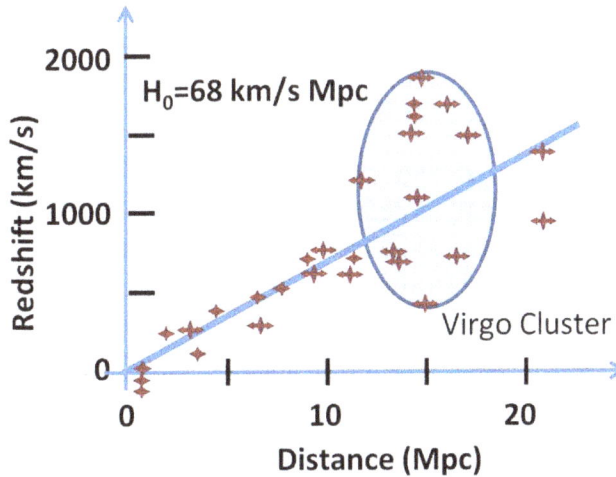

Fit of redshift velocities to Hubble's law. Various estimates for the Hubble constant exist. The HST Key H_0 Group fitted type Ia supernovae for redshifts between 0.01 and 0.1 to find that $H_0 = 71 \pm 2$ (statistical) ± 6 (systematic) km s^{-1}Mpc^{-1}, while Sandage *et al.* find $H_0 = 62.3 \pm 1.3$ (statistical) ± 5 (systematic) km s^{-1}Mpc^{-1}.

At the time of discovery and development of Hubble's law, it was acceptable to explain redshift phenomenon as a Doppler shift in the context of special relativity, and use the Doppler formula to associate redshift z with velocity. Today, in the context of general relativity, velocity between distant objects depends on the choice of coordinates used, and therefore, the redshift can be equally described as a Doppler shift or a cosmological shift (or gravitational) due to the expanding space, or some combination of the two.

Hubble Diagram

Hubble's law can be easily depicted in a "Hubble Diagram" in which the velocity (assumed approximately proportional to the redshift) of an object is plotted with respect to its distance from the observer. A straight line of positive slope on this diagram is the visual depiction of Hubble's law.

Cosmological Constant Abandoned

After Hubble's discovery was published, Albert Einstein abandoned his work on the cosmological constant, which he had designed to modify his equations of general relativity to allow them to produce a static solution, which he thought was the correct state of the universe. The Einstein equations in their simplest form model generally either an expanding or contracting universe, so Einstein's cosmological constant was artificially created to counter the expansion or contraction to get a perfect static and flat universe. After Hubble's discovery that the Universe was, in fact, expanding, Einstein called his faulty assumption that the Universe is static his "biggest mistake". On its own, general relativity could predict the expansion of the Universe, which (through observations such as the bending of light by large masses, or the precession of the orbit of Mercury) could be experimentally observed and compared to his theoretical calculations using particular solutions of the equations he had originally formulated.

In 1931, Einstein made a trip to Mount Wilson to thank Hubble for providing the observational basis for modern cosmology.

The cosmological constant has regained attention in recent decades as a hypothesis for dark energy.

Expansion Velocity vs Relative Velocity

In using Hubble's law to determine distances, only the velocity due to the expansion of the Universe can be used. Since gravitationally interacting galaxies move relative to each other independent of the expansion of the Universe, these relative velocities, called peculiar velocities, need to be accounted for in the application of Hubble's law.

The Finger of God effect is one result of this phenomenon. In systems that are gravitationally bound, such as galaxies or our planetary system, the expansion of space is a much weaker effect than the attractive force of gravity.

Exponentiality

While current evidence suggests that the expansion of the universe is accelerating, Hubble's Law implies that *all* derivatives of the expansion of the universe are increasing. This is readily seen as the comoving distance, D, is proportional to its own time derivative, v. Solving the relation for time yields

$$t = \int_{D_i}^{D} \frac{dD}{H_0 \cdot D},$$

where D_i is the distance at which a galaxy is first measured at and t is the time since said measurement, thus the velocity of a galaxy can be expressed as

$$v = \frac{D - D_i}{t}.$$

Solving the integral subsequently yields $D = e^{H_0 t} D_i$, with the implication that the comoving distance between two galaxies increases exponentially as time goes on. This lines up with current observations, however; it has not been shown for derivatives of distance above acceleration.

The above relations only hold so long as the Hubble parameter, H_0, is constant; which it is presumed not to be. In order for the acceleration of the expansion of the universe to stop, however; the Hubble parameter must be inversely proportional to time. That is, $H_0 = \frac{k}{t}$ for some constant k, giving us a formula for the comoving distance of a galaxy $D = e^k D_i$. Importantly, D is a constant in this scenario, leading to a static universe. As a result, the exponential expansion of the universe and its time derivatives are dependent on $\frac{dH_0}{dt}$, or how quickly the Hubble parameter is changing over time.

Idealized Hubble's Law

The mathematical derivation of an idealized Hubble's Law for a uniformly expanding universe is a fairly elementary theorem of geometry in 3-dimensional Cartesian/Newtonian coordinate space, which, considered as a metric space, is entirely homogeneous and isotropic (properties do not vary with location or direction). Simply stated the theorem is this:

Any two points which are moving away from the origin, each along straight lines and with speed proportional to distance from the origin, will be moving away from each other with a speed proportional to their distance apart.

In fact this applies to non-Cartesian spaces as long as they are locally homogeneous and isotropic; specifically to the negatively and positively curved spaces frequently considered as cosmological models.

An observation stemming from this theorem is that seeing objects recede from us on Earth is not an indication that Earth is near to a center from which the expansion is occurring, but rather that *every* observer in an expanding universe will see objects receding from them.

Ultimate Fate and Age of the Universe

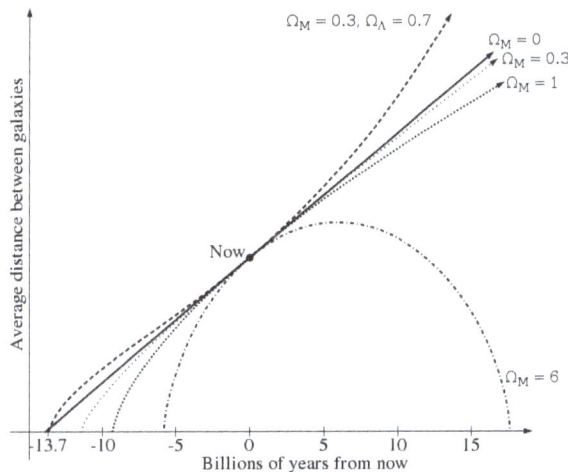

The age and ultimate fate of the universe can be determined by measuring
the Hubble constant today and extrapolating with the observed value of the deceleration parameter,
uniquely characterized by values of density parameters (Ω_M for matter and Ω_Λ for dark energy).

A "closed universe" with $\Omega_M > 1$ and $\Omega_\Lambda = 0$ comes to an end in a Big Crunch and is considerably younger than its Hubble age. An "open universe" with $\Omega_M \le 1$ and $\Omega_\Lambda = 0$ expands forever and has an age that is closer to its Hubble age. For the accelerating universe with nonzero Ω_Λ that we inhabit, the age of the universe is coincidentally very close to the Hubble age.

The value of the Hubble parameter changes over time, either increasing or decreasing depending on the value of the so-called deceleration parameter q, which is defined by

$$q = -\left(1 + \frac{\dot{H}}{H^2}\right).$$

In a universe with a deceleration parameter equal to zero, it follows that $H = 1/t$, where t is the time since the Big Bang. A non-zero, time-dependent value of q simply requires integration of the Friedmann equations backwards from the present time to the time when the comoving horizon size was zero.

It was long thought that q was positive, indicating that the expansion is slowing down due to gravitational attraction. This would imply an age of the Universe less than $1/H$ (which is about 14 billion years). For instance, a value for q of $1/2$ (once favoured by most theorists) would give the age of the Universe as $2/(3H)$. The discovery in 1998 that q is apparently negative means that the Universe could actually be older than $1/H$. However, estimates of the age of the universe are very close to $1/H$.

Olbers' Paradox

The expansion of space summarized by the Big Bang interpretation of Hubble's Law is relevant to the old conundrum known as Olbers' paradox: if the Universe were infinite, static, and filled with a uniform distribution of stars, then every line of sight in the sky would end on a star, and the sky would be as bright as the surface of a star. However, the night sky is largely dark. Since the 17th century, astronomers and other thinkers have proposed many possible ways to resolve this paradox, but the currently accepted resolution depends in part on the Big Bang theory and in part on the Hubble expansion. In a universe that exists for a finite amount of time, only the light of a finite number of stars has had a chance to reach us yet, and the paradox is resolved. Additionally, in an expanding universe, distant objects recede from us, which causes the light emanating from them to be redshifted and diminished in brightness.

Dimensionless Hubble Parameter

Instead of working with Hubble's constant, a common practice is to introduce the dimensionless Hubble parameter, usually denoted by h, and to write the Hubble's parameter H_0 as $h \times 100$ km s^{-1} Mpc^{-1}, all the uncertainty relative of the value of H_0 being then relegated to h. If a subscript is presented after h, it refers to the value of h used in that text's preceding calculation, and is equal to $H_0 / 100$. Currently $h = 0.678$, which can be represented as $h_{0.678}$. This should not be confused with the dimensionless value of Hubble's constant, usually expressed in terms of Planck units, with current value of $H_0 \times t_P = 1.18 \times 10^{-61}$.

Determining the Hubble Constant

The value of the Hubble constant is estimated by measuring the redshift of distant galaxies and then determining the distances to the same galaxies (by some other method than Hubble's law). Uncertainties in the physical assumptions used to determine these distances have caused varying estimates of the Hubble constant.

The observations of astronomer Walter Baade led him to define distinct "populations" for stars (Population I and Population II). The same observations led him to discover that there are two types of Cepheid variable stars. Using this discovery he recalculated the size of the known universe, doubling the previous calculation made by Hubble in 1929. He announced this finding to considerable astonishment at the 1952 meeting of the International Astronomical Union in Rome.

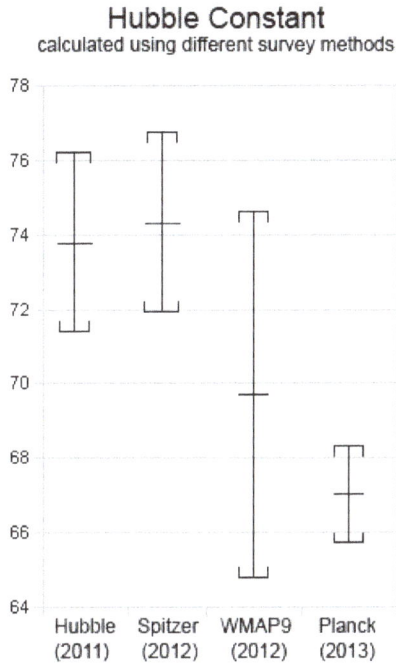

Value of the Hubble Constant including measurement uncertainty for recent surveys.

Earlier Measurement and Discussion Approaches

For most of the second half of the 20th century the value of H_0 was estimated to be between 50 and 90 (km/s)/Mpc.

The value of the Hubble constant was the topic of a long and rather bitter controversy between Gérard de Vaucouleurs, who claimed the value was around 100, and Allan Sandage, who claimed the value was near 50. In 1996, a debate moderated by John Bahcall between Sidney van den Bergh and Gustav Tammann was held in similar fashion to the earlier Shapley-Curtis debate over these two competing values.

This previously wide variance in estimates was partially resolved with the introduction of the ΛCDM model of the Universe in the late 1990s. With the ΛCDM model observations of high-redshift clusters at X-ray and microwave wavelengths using the Sunyaev-Zel'dovich effect, measurements of anisotropies in the cosmic microwave background radiation, and optical surveys all gave a value of around 70 for the constant.

More recent measurements from the Planck mission indicate a lower value of around 67; although, even more recently, in February 2018, a higher value of 73 has been determined using an improved procedure involving the Hubble Space Telescope.

Acceleration of the Expansion

A value for q measured from standard candle observations of Type Ia supernovae, which was determined in 1998 to be negative, surprised many astronomers with the implication that the expansion of the Universe is currently "accelerating".

Derivation of the Hubble Parameter

Start with the Friedmann equation:

$$H^2 \equiv \left(\frac{\dot{a}}{a}\right)^2 = \frac{8\pi G}{3}\rho - \frac{kc^2}{a^2} + \frac{\Lambda c^2}{3},$$

where H is the Hubble parameter, a is the scale factor, G is the gravitational constant, k is the normalised spatial curvature of the Universe and equal to -1, 0, or 1, and Λ is the cosmological constant.

Matter-dominated Universe

If the Universe is matter-dominated, then the mass density of the Universe ρ can just be taken to include matter so

$$\rho = \rho_m(a) = \frac{\rho_{m_0}}{a^3},$$

where ρ_{m_0} is the density of matter today. We know for nonrelativistic particles that their mass density decreases proportional to the inverse volume of the Universe, so the equation above must be true. We can also define

$$\rho_c = \frac{3H^2}{8\pi G};$$

$$\Omega_m \equiv \frac{\rho_{m_0}}{\rho_c} = \frac{8\pi G}{3H_0^2}\rho_{m_0};$$

therefore:

$$\rho = \frac{\rho_c \Omega_m}{a^3}.$$

Also, by definition,

$$\Omega_k \equiv \frac{-kc^2}{(a_0 H_0)^2}$$

$$\Omega_\Lambda \equiv \frac{\Lambda c^2}{3H_0^2},$$

where the subscript nought refers to the values today, and $a_0 = 1$. Substituting all of this into the Friedmann equation at the start of this section and replacing a with $a = 1/(1+z)$ gives

$$H^2(z) = H_0^2 \left(\Omega_m (1+z)^3 + \Omega_k (1+z)^2 + \Omega_\Lambda \right).$$

Matter- and Dark Energy-dominated Universe

If the Universe is both matter-dominated and dark energy- dominated, then the above equation for the Hubble parameter will also be a function of the equation of state of dark energy. So now:

$$\rho = \rho_m(a) + \rho_{de}(a),$$

where ρ_{de} is the mass density of the dark energy. By definition, an equation of state in cosmology is $P = w\rho c^2$, , and if this is substituted into the fluid equation, which describes how the mass density of the Universe evolves with time, then

$$\dot\rho + 3\frac{\dot a}{a}\left(\rho + \frac{P}{c^2}\right) = 0;$$

$$\frac{d\rho}{\rho} = -3\frac{da}{a}(1+w).$$

If w is constant, then

$$\ln \rho = -3(1+w)\ln a;$$

implying:

$$\rho = a^{-3(1+w)}.$$

Therefore, for dark energy with a constant equation of state w, $\rho_{de}(a) = \rho_{de0}a^{-3(1+w)}$. If this is substituted into the Friedman equation in a similar way as before, but this time set $k = 0$, which assumes a spatially flat universe, then

$$H^2(z) = H_0^2 \left(\Omega_m(1+z)^3 + \Omega_{de}(1+z)^{3(1+w)} \right).$$

If the dark energy derives from a cosmological constant such as that introduced by Einstein, it can be shown that $w = -1$. The equation then reduces to the last equation in the matter dominated universe section, with Ω_k set to zero. In that case the initial dark energy density ρ_{de0} is given by

$$\rho_{de0} = \frac{\Lambda c^2}{8\pi G} \text{ and } \Omega_{de} = \Omega_\Lambda.$$

If dark energy does not have a constant equation-of-state w, then

$$\rho_{de}(a) = \rho_{de0}e^{-3\int \frac{da}{a}(1+w(a))}$$

and to solve this, $w(a)$ must be parametrized, for example if $w(a) = w_0 + w_a(1-a)$, giving

$$H^2(z) = H_0^2 \left(\Omega_m a^{-3} + \Omega_{de} a^{-3(1+w_0+w_a)} e^{-3w_a(1-a)} \right).$$

Other ingredients have been formulated recently.

Units Derived from the Hubble Constant

Hubble Time

The Hubble constant H_0 has units of inverse time; the Hubble time t_H is simply defined as the inverse of the Hubble constant, i.e.

$$t_H \equiv \frac{1}{H_0} = \frac{1}{67.8 (km/s)/Mpc} = 4.55 \cdot 10^{17} s = 14.4 \text{ billion years.}$$

This is slightly different from the age of the universe which is approximately 13.8 billion years. The Hubble time is the age it would have had if the expansion had been linear, and it is different from the real age of the universe because the expansion is not linear; they are related by a dimensionless factor which depends on the mass-energy content of the universe, which is around 0.96 in the standard Lambda-CDM model.

We currently appear to be approaching a period where the expansion of the universe is exponential due to the increasing dominance of vacuum energy. In this regime, the Hubble parameter is constant, and the universe grows by a factor e each Hubble time:

$$H \equiv \frac{\dot{a}}{a} = \text{constant} \quad \Rightarrow \quad a \propto e^{Ht} = e^{\frac{t}{t_H}}$$

Over long periods of time, the dynamics are complicated by general relativity, dark energy, inflation, etc., as explained above.

Hubble Length

The Hubble length or Hubble distance is a unit of distance in cosmology, defined as cH_0^{-1} — the speed of light multiplied by the Hubble time. It is equivalent to 4,550 million parsecs or 14.4 billion light years. (The numerical value of the Hubble length in light years is, by definition, equal to that of the Hubble time in years.) The Hubble distance would be the distance between the Earth and the galaxies which are *currently* receding from us at the speed of light, as can be seen by substituting $D = cH_0^{-1}$ into the equation for Hubble's law, $v = H_0 D$.

Hubble Volume

The Hubble volume is sometimes defined as a volume of the Universe with a comoving size of cH_0^{-1} The exact definition varies: it is sometimes defined as the volume of a sphere with radius cH_0^{-1} or alternatively, a cube of side cH_0^{-1} Some cosmologists even use the term Hubble volume to refer to the volume of the observable universe, although this has a radius approximately three times larger.

Cosmological Perturbation Theory

Cosmological perturbation theory is a theory which explains how such structures can be formed from very small inhomogeneities in an otherwise homogeneous universe. One assumes the universe to be homogeneous and isotropic to the zeroth order, i.e. that it obeys the Friedmann-Robertson-Walker line element to this order,

$$ds^2 = a^2(\eta)\left(d\eta^2 - \delta_{ij}dx^i dx^j\right)$$

where η is conformal time and we have used units so that the speed of light $c = 1$. The conformal time relates to the usual comoving time, t, in the following way

$$a^2(\eta)d\eta^2 = dt^2 \quad \Rightarrow \quad t = \int_0^\eta a(\eta')d\eta'$$

In the expression above, we have assumed a flat FRW universe. The reason for us not including the open and the closed FRW universes, is that recent experimental cosmological observations have pretty much confirmed that the geometry of the Universe is indeed flat to a high degree of accuracy. Data from the boomerang balloon experiment, and also from the more recent WMAP satellite both support this conclusion.

Inhomogeneities are introduced as a first order perturbation to this metric, $\delta g_{\mu\nu}$ Thus, the physical, inhomogeneous line element can be written as

$$ds^2 = \left({}^{(0)}g_{\mu\nu} + \delta g_{\mu\nu}\right)dx^\mu dx^\nu$$

where ${}^{(0)}g_{\mu\nu}$ is the flat FRW metric.

Gauge-invariant Perturbation Theory

The gauge-invariant perturbation theory is based on developments by Bardeen (1980), Kodama and Sasaki (1984) building on the work of Lifshitz (1946). This is the standard approach to perturbation theory of general relativity for cosmology. This approach is widely used for the computation of anisotropies in the cosmic microwave background radiation as part of the physical cosmology program and focuses on predictions arising from linearisations that preserve gauge invariance with respect to Friedmann-Lemaître-Robertson-Walker (FLRW) models. This approach draws heavily on the use of Newtonian like analogue and usually has as it starting point the FRW background around which perturbations are developed. The approach is non-local and coordinate dependent but gauge invariant as the resulting linear framework is built from a specified family of background hyper-surfaces which are linked by gauge preserving mappings to foliate the space-time. Although intuitive this approach does not deal well with the nonlinearities natural to general relativity.

1+3 Covariant Gauge-invariant Perturbation Theory

In relativistic cosmology using the Lagrangian threading dynamics of Ehlers (1971) and Ellis (1971) it is usual to use the gauge-invariant covariant perturbation theory developed by Hawking

(1966) and Ellis and Bruni (1989). Here rather than starting with a background and perturbing away from that background one starts with full general relativity and systematically reduces the theory down to one that is linear around a particular background. The approach is local and both covariant as well as gauge invariant but can be non-linear because the approach is built around the local comoving observer frame which is used to thread the entire space-time. This approach to perturbation theory produces differential equations that are of just the right order needed to describe the true physical degrees of freedom and as such no non-physical gauge modes exist. It is usual to express the theory in a coordinate free manner. For applications of kinetic theory, because one is required to use the full tangent bundle, it becomes convenient to use the tetrad formulation of relativistic cosmology. The application of this approach to the computation of anisotropies in cosmic microwave background radiation requires the linearization of the full relativistic kinetic theory developed by Thorne (1980) and Ellis, Matravers and Treciokas (1983).

Gauge Freedom and Frame Fixing

In relativistic cosmology there is a freedom associated with the choice of threading frame, this frame choice is distinct from choice associated with coordinates. Picking this frame is equivalent to fixing the choice of timelike world lines mapped into each other, this reduces the gauge freedom it does not fix the gauge but the theory remains gauge invariant under the remaining gauge freedoms. In order to fix the gauge a specification of correspondences between the time surfaces in the real universe (perturbed) and the background universe are required along with the correspondences between points on the initial spacelike surfaces in the background and in the real universe. This is the link between the gauge-invariant perturbation theory and the gauge-invariant covariant perturbation theory. Gauge invariance is only guaranteed if the choice of frame coincides exactly with that of the background; usually this is trivial to ensure because physical frames have this property.

Newtonian-like Equations

Newtonian-like equations emerge from perturbative general relativity with the choice of the Newtonian gauge; the Newtonian gauge provides the direct link between the variables typically used in the gauge-invariant perturbation theory and those arising from the more general gauge-invariant covariant perturbation theory.

Structure Formation

By *structure formation* we mean the generation and evolution of this inhomogeneity. We are here interested in distance scales from galaxy size to the size of the whole observable universe. The structure is manifested in the existence of galaxies and in their uneven distribution, their clustering. This is the obvious inhomogeneity, but we believe it reflects a density inhomogeneity also in other, nonluminous, components of the universe, especially the *cold dark matter*. We understand today that the structure has formed by gravitational amplification of a small primordial inhomogeneity. There are thus two parts to the theory of structure formation:

1) The generation of this primordial inhomogeneity, "the seeds of galaxies". This is the more speculative part of structure formation theory. We cannot claim that we know how this primordial inhomogeneity came about, but we have a good candidate theory, inflation, whose predictions agree with the present observational data, and can be tested more thoroughly by future observations. In the inflation theory, the structure originates from quantum fluctuations of the inflaton field ' near the time the scale in question exits the horizon.

2) The growth of this small inhomogeneity into the present observable structure of the universe. This part is less speculative, since we have a well established theory of gravity, general relativity. However, there is uncertainty in this part too, since we do not know the precise nature of the dominant components to the energy density of the universe, the dark matter and the dark energy. The gravitational growth depends on the equations of state and the streaming lengths (particle mean free path between interactions) of these density components. Besides gravity, the growth is affected by pressure forces.

We shall start our discussion from the second part. The basic tool here is firstorder perturbation theory (also called linear perturbation theory). This means that we write all our inhomogeneous quantities as a sum of the background value, corresponding to the homogeneous and isotropic model, and a perturbation, the deviation from the background value. For example, for the energy density we write

$$\rho(t,x) = \bar{\rho}(t) + \delta\rho(t,x).$$

We then assume that the perturbation is small, so it is a reasonable approximation to drop from our equations all those terms which contain a product of two or more perturbations. The remaining equation will then contain only terms which are either zeroth order, i.e., contain only background quantities, or first order, i.e., contain exactly one perturbation. If we kept only the zeroth order parts, we would be back to the equations of the homogenous and isotropic universe. Subtracting these from our equations we arrive at the perturbation equations where every term is first-order in the perturbation quantities, i.e., it is a linear equation for them. This makes the equations easy to handle, we can, e.g., Fourier transform them.

As we discovered in our discussion of inflation, the different cosmological distance scales first exit the horizon during inflation, then enter the horizon during various epochs of the later history. Perturbations at subhorizon scales can be treated with Newtonian perturbation theory (possibly slightly generalized to include high-pressure matter), but scales which are close to horizon size or superhorizon require relativistic perturbation theory, which is based on general relativity.

Under present models, the structure of the visible universe was formed in the following stages:

Very Early Universe

In this stage, some mechanism, such as cosmic inflation, was responsible for establishing the initial conditions of the universe: homogeneity, isotropy, and flatness. Cosmic inflation also would have amplified minute quantum fluctuations (pre-inflation) into slight density ripples of overdensity and underdensity (post-inflation).

Growth of Structure

The early universe was dominated by radiation; in this case density fluctuations larger than the cosmic horizon grow proportional to the scale factor, as the gravitational potential fluctuations remain constant. Structures smaller than the horizon remained essentially frozen due to radiation domination impeding growth. As the universe expanded, the density of radiation drops faster than matter (due to redshifting of photon energy); this led to a crossover called matter-radiation equality at ~ 50,000 years after the Big Bang. After this all dark matter ripples could grow freely, forming seeds into which the baryons could later fall. The size of the universe at this epoch forms a turnover in the matter power spectrum which can be measured in large redshift surveys.

Recombination

The universe was dominated by radiation for most of this stage, and due to the intense heat and radiation, the primordial hydrogen and helium were fully ionized into nuclei and free electrons. In this hot and dense situation, the radiation (photons) could not travel far before Thomson scattering off an electron. The universe was very hot and dense, but expanding rapidly and therefore cooling. Finally, at a little less than 400,000 years after the 'bang', it become cool enough (around 3000 K) for the protons to capture negatively charged electrons, forming neutral hydrogen atoms. (Helium atoms formed somewhat earlier due to their larger binding energy). Once nearly all the charged particles were bound in neutral atoms, the photons no longer interacted with them and were free to propagate for the next 13.8 billion years; we currently detect those photons redshifted by a factor 1090 down to 2.725 K as the Cosmic Microwave Background Radiation (CMB) filling today's universe. Several remarkable space-based missions (COBE, WMAP, Planck), have detected very slight variations in the density and temperature of the CMB. These variations were subtle, and the CMB appears very nearly uniformly the same in every direction. However, the slight temperature variations of order a few parts in 100,000 are of enormous importance, for they essentially were early "seeds" from which all subsequent complex structures in the universe ultimately developed.

The theory of what happened after the universe's first 400,000 years is one of hierarchical structure formation: the smaller gravitationally bound structures such as matter peaks containing the first stars and stellar clusters formed first, and these subsequently merged with gas and dark matter to form galaxies, followed by groups, clusters and superclusters of galaxies.

Very Early Universe

The very early universe is still a poorly understood epoch, from the viewpoint of fundamental physics. The prevailing theory, cosmic inflation, does a good job explaining the observed flatness, homogeneity and isotropy of the universe, as well as the absence of exotic relic particles (such as magnetic monopoles). Another prediction borne out by observation is that tiny perturbations in the primordial universe seed the later formation of structure. These fluctuations, while they form the foundation for all structure, appear most clearly as tiny temperature fluctuations at one part in 100,000. (To put this in perspective, the same level of fluctuations on a topographic map of the United States would show no feature taller than a few centimeters.) These fluctuations are critical, because they provide the seeds from which the largest structures can grow and eventually collapse to form galaxies and stars. COBE (Cosmic Background Explorer) provided the first detection of the intrinsic fluctuations in the cosmic microwave background radiation in the 1990s.

These perturbations are thought to have a very specific character: they form a Gaussian random field whose covariance function is diagonal and nearly scale-invariant. Observed fluctuations appear to have exactly this form, and in addition the *spectral index* measured by WMAP—the spectral index measures the deviation from a scale-invariant (or Harrison-Zel'dovich) spectrum—is very nearly the value predicted by the simplest and most robust models of inflation. Another important property of the primordial perturbations, that they are adiabatic (or isentropic between the various kinds of matter that compose the universe), is predicted by cosmic inflation and has been confirmed by observations.

Other theories of the very early universe have been proposed that are claimed to make similar predictions, such as the brane gas cosmology, cyclic model, pre-big bang model and holographic universe, but they remain nascent and are not widely accepted. Some theories, such as cosmic strings, have largely been refuted by increasingly precise data.

The Horizon Problem

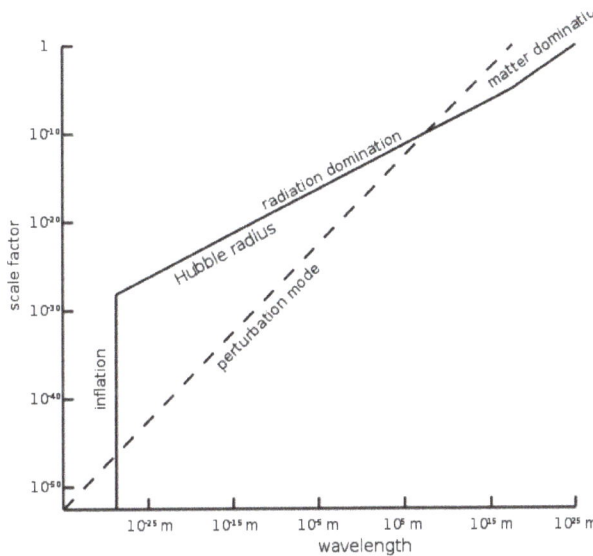

The physical size of the Hubble radius (solid line) as a function of the scale factor of the universe. The physical wavelength of a perturbation mode (dashed line) is shown as well.

The plot illustrates how the perturbation mode exits the horizon during cosmic inflation in order to reenter during radiation domination. If cosmic inflation never happened, and radiation domination continued back until a gravitational singularity, then the mode would never have exited the horizon in the very early universe.

An important concept in structure formation is the notion of the Hubble radius, often called simply the *horizon*, as it is closely related to the particle horizon. The Hubble radius, which is related to the Hubble parameter H as $R = c / H$, where c is the speed of light, defines, roughly speaking, the volume of the nearby universe that has recently (in the last expansion time) been in causal contact with an observer. Since the universe is continually expanding, its energy density is continually decreasing (in the absence of truly exotic matter such as phantom energy). The Friedmann equation relates the energy density of the universe to the Hubble parameter and shows that the Hubble radius is continually increasing.

The horizon problem of big bang cosmology says that, without inflation, perturbations were never in causal contact before they entered the horizon and thus the homogeneity and isotropy of, for example, the large scale galaxy distributions cannot be explained. This is because, in an ordinary Friedmann–Lemaître–Robertson–Walker cosmology, the Hubble radius increases more rapidly than space expands, so perturbations only enter the Hubble radius, and are not pushed out by the expansion. This paradox is resolved by cosmic inflation, which suggests that during a phase of rapid expansion in the early universe the Hubble radius was nearly constant. Thus, large scale isotropy is due to quantum fluctuations produced during cosmic inflation that are pushed outside the horizon.

Primordial Plasma

The end of inflation is called reheating, when the inflation particles decay into a hot, thermal plasma of other particles. In this epoch, the energy content of the universe is entirely radiation, with standard model particles having relativistic velocities. As the plasma cools, baryogenesis and leptogenesis are thought to occur, as the quark–gluon plasma cools, electroweak symmetry breaking occurs and the universe becomes principally composed of ordinary protons, neutrons and electrons. As the universe cools further, big bang nucleosynthesis occurs and small quantities of deuterium, helium and lithium nuclei are created. As the universe cools and expands, the energy in photons begins to redshift away, particles become non-relativistic and ordinary matter begins to dominate the universe. Eventually, atoms begin to form as free electrons bind to nuclei. This suppresses Thomson scattering of photons. Combined with the rarefaction of the universe (and consequent increase in the mean free path of photons), this makes the universe transparent and the cosmic microwave background is emitted at recombination (the *surface of last scattering*).

Acoustic Oscillations

The primordial plasma would have had very slight overdensities of matter, thought to have derived from the enlargement of quantum fluctuations during inflation. Whatever the source, these overdensities gravitationally attract matter. But the intense heat of the near constant photon-matter interactions of this epoch rather forcefully seeks thermal equilibrium, which creates a large amount of outward pressure. These counteracting forces of gravity and pressure create oscillations, analogous to sound waves created in air by pressure differences.

These perturbations are important, as they are responsible for the subtle physics that result in the cosmic microwave background anisotropy. In this epoch, the amplitude of perturbations that enter the horizon oscillate sinusoidally, with dense regions becoming more rarefied and then becoming dense again, with a frequency which is related to the size of the perturbation. If the perturbation oscillates an integral or half-integral number of times between coming into the horizon and recombination, it appears as an acoustic peak of the cosmic microwave background anisotropy. (A half-oscillation, in which a dense region becomes a rarefied region or vice versa, appears as a peak because the anisotropy is displayed as a *power spectrum*, so underdensities contribute to the power just as much as overdensities.) The physics that determines the detailed peak structure of the microwave background is complicated, but these oscillations provide the essence.

Linear Structure

One of the key realizations made by cosmologists in the 1970s and 1980s was that the majority of the matter content of the universe was composed not of atoms, but rather a mysterious form of matter known as dark matter. Dark matter interacts through the force of gravity, but it is not composed of baryons, and it is known with very high accuracy that it does not emit or absorb radiation. It may be composed of particles that interact through the weak interaction, such as neutrinos, but it cannot be composed entirely of the three known kinds of neutrinos (although some have suggested it is a sterile neutrino). Recent evidence indicates that there are about five times as much dark matter as baryonic matter, and thus the dynamics of the universe in this epoch are dominated by dark matter.

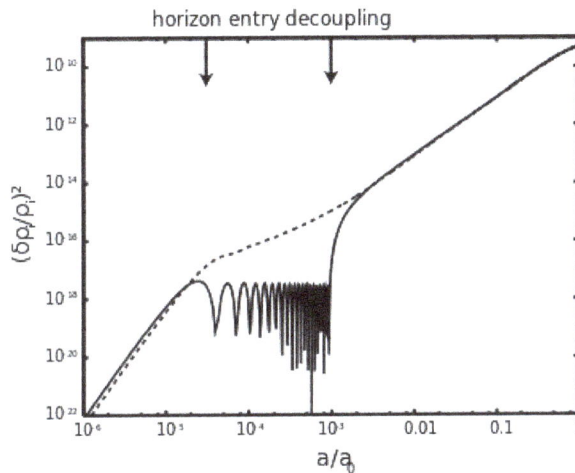

Between entering the horizon and decoupling, the dark matter perturbation (dashed line) grows logarithmically, before the growth accelerates in matter domination. On the other hand, between entering the horizon and decoupling, the perturbation in the baryon-photon fluid (solid line) oscillates rapidly.

After decoupling, it grows rapidly to match the dominant matter perturbation, the dark matter mode. Dark matter plays a crucial role in structure formation because it feels only the force of gravity: the gravitational Jeans instability which allows compact structures to form is not opposed by any force, such as radiation pressure. As a result, dark matter begins to collapse into a complex network of dark matter halos well before ordinary matter, which is impeded by pressure forces. Without dark matter, the epoch of galaxy formation would occur substantially later in the universe than is observed.

The physics of structure formation in this epoch is particularly simple, as dark matter perturbations with different wavelengths evolve independently. As the Hubble radius grows in the expanding universe, it encompasses larger and larger disturbances. During matter domination, all causal dark matter perturbations grow through gravitational clustering. However, the shorter-wavelength perturbations that are included during radiation domination have their growth retarded until matter domination. At this stage, luminous, baryonic matter is expected to mirror the evolution of the dark matter simply, and their distributions should closely trace one another.

It is a simple matter to calculate this "linear power spectrum" and, as a tool for cosmology, it is of comparable importance to the cosmic microwave background. Galaxy surveys have measured the power spectrum, such as the Sloan Digital Sky Survey, and by surveys of the Lyman-α forest. Since these studies observe radiation emitted from galaxies and quasars, they do not directly measure the dark matter, but the large-scale distribution of galaxies (and of absorption lines in the Lyman-α forest) is expected to mirror the distribution of dark matter closely. This depends on the fact that galaxies will be larger and more numerous in denser parts of the universe, whereas they will be comparatively scarce in rarefied regions.

Nonlinear Structure

When the perturbations have grown sufficiently, a small region might become substantially denser than the mean density of the universe. At this point, the physics involved becomes substantially more complicated. When the deviations from homogeneity are small, the dark matter may be treated as a pressureless fluid and evolves by very simple equations. In regions which are significantly denser than the background, the full Newtonian theory of gravity must be included. (The Newtonian theory is appropriate because the masses involved are much less than those required to form a black hole, and the speed of gravity may be ignored as the light-crossing time for the structure is still smaller than the characteristic dynamical time.) One sign that the linear and fluid approximations become invalid is that dark matter starts to form caustics in which the trajectories of adjacent particles cross, or particles start to form orbits. These dynamics are best understood using N-body simulations (although a variety of semi-analytic schemes, such as the Press–Schechter formalism, can be used in some cases). While in principle these simulations are quite simple, in practice they are tough to implement, as they require simulating millions or even billions of particles. Moreover, despite the large number of particles, each particle typically weighs 10^9 solar masses and discretization effects may become significant. The largest such simulation as of 2005 is the Millennium simulation.

The result of N-body simulations suggests that the universe is composed largely of voids, whose densities might be as low as one-tenth the cosmological mean. The matter condenses in large filaments and haloes which have an intricate web-like structure. These form galaxy groups, clusters and superclusters. While the simulations appear to agree broadly with observations, their interpretation is complicated by the understanding of how dense accumulations of dark matter spur galaxy formation. In particular, many more small haloes form than we see in astronomical observations as dwarf galaxies and globular clusters. This is known as the galaxy bias problem, and a variety of explanations have been proposed. Most account for it as an effect in the complicated physics of galaxy formation, but some have suggested that it is a problem with our model of dark matter and that some effect, such as warm dark matter, prevents the formation of the smallest haloes.

Gas Evolution

The final stage in evolution comes when baryons condense in the centres of galaxy haloes to form galaxies, stars and quasars. Dark matter greatly accelerates the formation of dense haloes. As dark matter does not have radiation pressure, the formation of smaller structures from dark matter is impossible. This is because dark matter cannot dissipate angular momentum, whereas ordinary baryonic matter can collapse to form dense objects by dissipating angular momentum through radiative cooling. Understanding these processes is an enormously difficult computational problem, because they can involve the physics of gravity, magnetohydrodynamics, atomic physics, nuclear reactions, turbulence and even general relativity. In most cases, it is not yet possible to perform simulations that can be compared quantitatively with observations, and the best that can be achieved are approximate simulations that illustrate the main qualitative features of a process such as a star formation.

Modelling Structure Formation

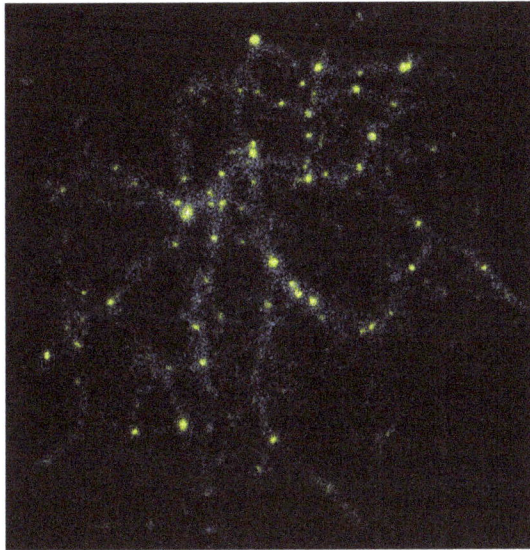

Snapshot from a computer simulation of large scale structure formation in a Lambda-CDM universe.

Cosmological Perturbations

Much of the difficulty, and many of the disputes, in understanding the large-scale structure of the universe can be resolved by better understanding the choice of gauge in general relativity. By the scalar-vector-tensor decomposition, the metric includes four scalar perturbations, two vector perturbations, and one tensor perturbation. Only the scalar perturbations are significant: the vectors are exponentially suppressed in the early universe, and the tensor mode makes only a small (but important) contribution in the form of primordial gravitational radiation and the B-modes of the cosmic microwave background polarization. Two of the four scalar modes may be removed by a physically meaningless coordinate transformation. Which modes are eliminated determine the infinite number of possible gauge fixings. The most popular gauge is Newtonian gauge (and the closely related conformal Newtonian gauge), in which the retained scalars are the Newtonian potentials Φ and Ψ, which correspond exactly to the Newtonian potential energy from Newtonian gravity. Many other gauges are used, including synchronous gauge, which can be an efficient gauge for numerical computation (it is used by CMBFAST). Each gauge still includes some unphysical degrees of freedom. There is a so-called gauge-invariant formalism, in which only gauge invariant combinations of variables are considered.

Inflation and Initial Conditions

The initial conditions for the universe are thought to arise from the scale invariant quantum mechanical fluctuations of cosmic inflation. The perturbation of the background energy density at a given point $\rho(\mathbf{x},t)$ in space is then given by an isotropic, homogeneous Gaussian random field of mean zero. This means that the spatial Fourier transform of $\rho - \hat{\rho}(\mathbf{k},t)$ has the following correlation functions

$$\langle \hat{\rho}(\mathbf{k},t)\hat{\rho}(\mathbf{k}',t) \rangle = f(k)\delta^{(3)}(\mathbf{k}-\mathbf{k}'),$$

where $\delta^{(3)}$ is the three-dimensional Dirac delta function and $k = |\mathbf{k}|$ is the length of $\mathbf{k}$. Moreover, the spectrum predicted by inflation is nearly scale invariant, which means

$$\langle \hat{\rho}(\mathbf{k},t)\hat{\rho}(\mathbf{k}',t) \rangle = k^{n_s-1}\delta^{(3)}(\mathbf{k}-\mathbf{k}'),$$

where $n_s - 1$ is a small number. Finally, the initial conditions are adiabatic or isentropic, which means that the fractional perturbation in the entropy of each species of particle is equal. The resulting predictions fit very well with observations, however there is a conceptual problem with the physical picture presented above. The quantum state from which the quantum fluctuations are extracted, is in fact completely homogeneous and isotropic, and thus it can not be argued that the quantum fluctuations represent the primordial inhomogeneities and anisotropies. The interpretation of quantum uncertainties in the value of the inflation field (which is what the so-called quantum fluctuations really are) as if they were statistical fluctuations in a Gaussian random field does not follow from the application of standard rules of quantum theory. The issue is sometimes presented in terms of the "quantum to classical transition", which is a confusing manner to refer to the problem at hand, as there are very few physicists, if any, that would argue that there is any entity that is truly classical at the fundamental level. In fact, the consideration of these issues brings us face to face with the so called measurement problem in quantum theory. If anything, the problem becomes exacerbated in the cosmological context, as the early universe contains no entities that might be taken as playing the role of "observers" or of "measuring devices", both of which are essential for the standard usage of quantum mechanics. The most popular posture among cosmologists, in this regard, is to rely on arguments based on decoherence and some form of "Many Worlds Interpretation" of quantum theory. There is an intense ongoing debate about the reasonableness of that posture .

Baryogenesis

In cosmology and particle physics, by *baryogenesis* one refers to the process by which an abundance of baryons is produced in the early observable universe, leading to the baryonic matter seen all around. Since in the standard model of particle physics matter and antimatter are *essentially* always produced symmetrically ("conservation of baryon number"), the topic of baryogenesis is to a large extent the study of how the matter/antimatter symmetry may be broken.

There is good evidence that there are no large regions of antimatter (Cohen, De Rujula, and Glashow, 1998). It was Andrei Sakharov (1967) who first suggested that the baryon density might not represent some sort of initial condition, but might be understandable in terms of microphysical laws. He listed three ingredients to such an understanding:

1. Baryon number violation must occur in the fundamental laws. At very early times, if baryon number violating interactions were in equilibrium, then the universe can be said to have "started" with zero baryon number. Starting with zero baryon number, baryon number violating interactions are obviously necessary if the universe is to end up with a non-zero asymmetry. As we will see, apart from the philosophical appeal of these ideas, the success of inflationary theory suggests that, shortly after the big bang, the baryon number was essentially zero.

2. CP-violation: If CP (the product of charge conjugation and parity) is conserved, every reaction which produces a particle will be accompanied by a reaction which produces its antiparticle at precisely the same rate, so no baryon number can be generated.

3. An Arrow of Time (Departure from Thermal Equilibrium): The universe, for much of its history, was very nearly in thermal equilibrium. The spectrum of the CMBR is the most perfect blackbody spectrum measured in nature. So the universe was certainly in thermal equilibrium 105 years after the big bang. The success of the theory of big bang nucleosynthesis (BBN) provides strong evidence that the universe was in equilibrium two-three minutes after the big bang. But if, through its early history, the universe was in thermal equilibrium, then even B and CP violating interactions could not produce a net asymmetry. One way to understand this is to recall that the CPT theorem assures strict equality of particle and antiparticle masses, so at thermal equilibrium, the densities of particles and antiparticles are equal. More precisely, since B is odd under CPT, its thermal average vanishes in an equilibrium situation. This can be generalized by saying that the universe must have an arrow of time.

Several mechanisms have been proposed to understand the baryon asymmetry:

1. GUT Baryogenesis. Grand Unified Theories unify the gauge interactions of the strong, weak and electromagnetic interactions in a single gauge group. They inevitably violate baryon number, and they have heavy particles, with mass of order $M_{GUT} \approx 10_{16}$ GeV, whose decays can provide a departure from equilibrium. The main objections to this possibility come from issues associated with inflation. While there does not exist a compelling microphysical model for inflation, in most models, the temperature of the universe after reheating is well below M_{GUT}. But even if it were very large, there would be another problem. Successful unification requires supersymmetry, which implies that the graviton has a spin-3/2 partner, called the gravitino. In most models for supersymmetry breaking, these particles have masses of order TeV, and are very long lived. Even though these particles are weakly interacting, too many gravitinos are produced unless the reheating temperature is well below the unification scale -- too low for GUT baryogenesis to occur.

2. Electroweak baryogenesis. The Standard Model satisfies all of the conditions for baryogenesis, but any baryon asymmetry produced is far too small to account for observations. In certain extensions of the Standard Model, it is possible to obtain an adequate asymmetry, but in most cases the allowed region of parameter space is very small.

3. Leptogenesis. The possibility that the weak interactions will convert some lepton number to baryon number means that if one produces a large lepton number at some stage, this will be processed into a net baryon and lepton number at the electroweak phase transition. The

observation of neutrino masses makes this idea highly plausible. Many but not all of the relevant parameters can be directly measured.

4. Production by coherent motion of scalar fields (the Affleck-Dine mechanism), which can be highly efficient, might well be operative if nature is supersymmetric.

GUTs Naturally Satisfy all Three of Sakharov's Conditions

Baryon number (B) violation is a hallmark of these theories: they typically contain gauge bosons and other fields which mediate B violating interactions such as proton decay.

CP violation is inevitable; necessarily, any model contains at least the Kobayashi- Maskawa (KM) mechanism for violating CP, and typically there are many new couplings which can violate CP.

Departure from equilibrium is associated with the dynamics of the massive, B violating fields. Typically one assumes that these fields are in equilibrium at temperatures well above the grand unification scale. As the temperature becomes comparable to their mass, the production rates of these particles fall below their rates of decay. Careful calculations in these models often lead to baryon densities compatible with what we observe.

Example: SU(5) GUT. Treat all quarks and leptons as left-handed fields. In a single generation of quarks and leptons one has the quark doublet Q, the singlet u-bar and d-bar antiquarks (their antiparticles are the right-handed quarks), and the lepton doublet, L. Then it is natural to identify the fields in the 5-bar as follows:

$$L = \begin{pmatrix} e \\ v \end{pmatrix} \qquad \bar{5}_i = \begin{pmatrix} \bar{d} \\ \bar{d} \\ \bar{d} \\ e \\ v \end{pmatrix}$$

The remaining quarks and leptons (e- and e+) are in a 10 of SU(5).

The gauge fields are in the 24 (adjoint) representation:

Color SU(3) $T = \begin{pmatrix} \frac{\lambda^a}{2} & 0 \\ 0 & 0 \end{pmatrix}$ Weak SU(2) $T = \begin{pmatrix} 0 & 0 \\ 0 & \frac{\sigma^t}{2} \end{pmatrix}$

The U(1) generator is

$$Y' = \frac{1}{\sqrt{60}} \begin{pmatrix} 2 & & & & \\ & 2 & & & \\ & & 2 & & \\ & & & -3 & \\ & & & & -3 \end{pmatrix}$$

SU(5) is a broken symmetry, and it can be broken by a scalar Higgs field proportional to Y'. The unbroken symmetries are generated by the operators that commute with Y', namely SU(3)xSU(2) xU(1). The vector bosons X that correspond to broken generators, for example

$$\begin{pmatrix} 0 & 0 & 1 & 0 \\ 0 & 0 & 0 & 0 \\ 0 & 0 & 0 & 0 \\ 1 & 0 & 0 & 0 \\ 0 & 0 & 0 & 0 \end{pmatrix}$$

gain mass ~1016 GeV by this GUT HIggs mechanism.

The X bosons carry color and electroweak quantum numbers and mediate processes which violate baryon number. For example, there is a coupling of the X bosons to a d-bar quark and an electron.

In the GUT picture of baryogenesis, it is usually assumed that at temperatures well above the GUT scale, the universe was in thermal equilibrium. As the temperature drops below the mass of the X bosons, the reactions which produce the X bosons are not sufficiently rapid to maintain equilibrium. The decays of the X bosons violate baryon number; they also violate CP. So all three conditions are readily met: B violation, CP violation, and departures from equilibrium.

CPT requires that the total decay rate of X is the same as that of its antiparticle X-bar. But it does not require equality of the decays to particular final states (partial widths). So starting with equal numbers of X and X-bar particles, there can be a slight asymmetry between the processes

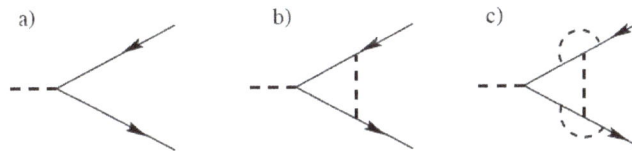

Interference between the tree-level (a) and one-loop (b) diagrams with complex Yukawa couplings can provide the requisite source of CP violation for GUT baryogenesis. In viable models, to avoid unwanted cancellations, one must often assume that the two scalars are different or go to higher loops (c)

$$X \rightarrow dL; X \rightarrow \overline{Q}\overline{u}$$

And

$$\overline{X} \rightarrow \overline{dL}; \overline{X} \rightarrow Qu$$

This can result in a slight excess of matter over antimatter. But reheating to T >1016 GeV after inflation will overproduce gravitinos -- so GUT baryogenesis is now disfavored.

Electroweak Baryogenesis

Below the electroweak scale of ~ 100 GeV, the sphaleron quantum tunneling process that violates B and L conservation (but preserves B - L) in the Standard Model is greatly suppressed, by ~ $\exp(2\pi/\alpha_w)$ ~ 10-65. But at T ~ 100 GeV this process can occur. It can satisfy all three Sakharov conditions, but it cannot produce a large enough B and L. However, it can easily convert L into a mixture of B and L (Leptogenesis).

When one quantizes the Standard Model, one finds that the baryon number current is not exactly conserved, but rather satisfies

$$\partial_\mu j_B^\mu = \frac{3}{16\pi^2} F_{\mu\nu}^a \tilde{F}_{\mu\nu}^a = \frac{3}{8\pi^2} \text{Tr}\, F_{\mu\nu} \tilde{F}_{\mu\nu.}$$

The same parity-violating term occurs in the divergence of the lepton number current, so the difference (the B - L current) is exactly conserved. The parityviolating term is a total divergence

$$\text{Tr}\, F_{\mu\nu} \tilde{F}_{\mu\nu.} = \partial_\mu K^\mu \qquad \text{where}\quad K^\mu = \varepsilon^{\mu\nu\rho\sigma}\, tr\left[F_{\nu\rho} A_\sigma + \frac{2}{3} A_\nu A_\rho A_\sigma \right], \text{So}$$

$$\tilde{j} = j_B^\mu - \frac{3g^2}{8\pi^2} K^\mu \quad \text{is conserved. In perturbation theory (i.e. Feynman diagrams)}$$

K^μ falls to zero rapidly at infinity, so B and L are conserved.

In abelian -- i.e. U(1) -- gauge theories, this is the end of the story. In non-abelian theories, however, there are non-perturbative field configurations, called instantons, which lead to violations of B and L. They correspond to calculation of a tunneling amplitude. To understand what the tunneling process is, one must consider more carefully the ground state of the field theory. Classically, the ground states are field configurations for which the energy vanishes. The trivial solution of this condition is

A = 0, where A is the vector potential, which is the only possibility in U(1). But a "pure gauge" is also a solution, where

$$\vec{A} = \frac{1}{i} g^1 \vec{\nabla} g,$$

where g is a gauge transformation matrix. There is a class of gauge transformations g, labeled by a discrete index n, which must also be considered. These have the form

$$g_n(\vec{x}) = e^{in} f^{(\vec{x})\hat{x}.r/2} \quad \text{where } f(x) \to 2\pi \text{ as } \vec{x} \to \infty, \text{and } f(\vec{x}) \to 0 \text{ as } \vec{x} \to 0.$$

The ground states are labeled by the index n. If we evaluate the integral of the current K^μ we obtain a quantity known as the Chern-Simons number

$$n_{CS} = \frac{1}{16\pi^2} \int d^3x\, K^0 = \frac{2/3}{16\pi^2} \int d^3x \in_{ijk} Tr\left(g^{-1}\partial_i g g^{-1}\partial_j g g^{-1}\partial_{kg} \right). \text{For } g = gn, n_{CS} = n.$$

Schematic Yang-Mills vacuum structure. At zero temperature, the instanton transitions between vacua with different Chern-Simons numbers are suppressed. At finite temperature, these transitions can proceed via sphalerons.

In tunneling processes which change the Chern-Simons number, because of the anomaly, the baryon and lepton numbers will change. The exponential suppression found in the instanton calculation is typical of tunneling processes, and in fact the instanton calculation is nothing but a field-theoretic WKB calculation. The probability that a single proton has decayed through this process in the history of the universe is infinitesimal. But this picture suggests that, at finite temperature, the rate should be larger. One can determine the height of the barrier separating configurations of different nCS by looking for the field configuration which corresponds to sitting on top of the barrier. This is a solution of the static equations of motion with finite energy. It is known as a "sphaleron". It follows that when the temperature is of order the ElectroWeak scale ~ 100 GeV, B and L violating (but B - L conserving) processes can proceed rapidly.

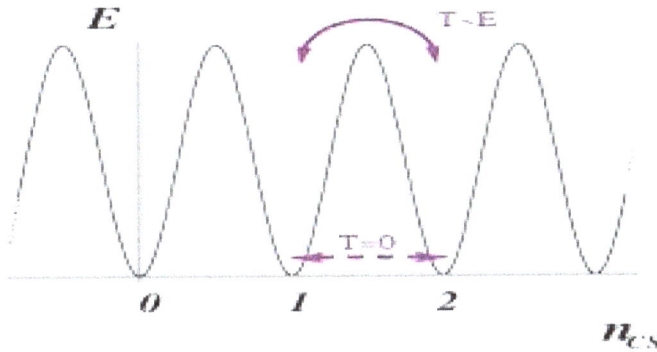

This result leads to three remarks:

1. If in the early universe, one creates baryon and lepton number, but no net B − L, B and L will subsequently be lost through sphaleron processes.

2. If one creates a net B − L (e.g. creates a lepton number) the sphaleron process will leave both baryon and lepton numbers comparable to the original B − L. This realization is crucial to the idea of Leptogenesis.

3. The Standard Model satisfies, by itself, all of the conditions for baryogenesis. However, detailed calculations show that in the Standard Model the size of the baryon and lepton numbers produced are much too small to be relevant for cosmology, both because the Higgs boson is more massive than ~ 80 GeV and because the CKM CP violation is much too small. In supersymmetric extensions of the Standard Model it is possible that a large enough matterantimatter asymmetry might be generated, but the parameter space for this is extremely small.

This leaves Leptogenesis and Affleck-Dine baryogenesis as the two most promising possibilities. What is exciting about each of these is that, if they are operative, they have consequences for experiments which will be performed at accelerators over the next few years.

Leptogenesis

There is now compelling experimental evidence that neutrinos have mass, both from solar and atmospheric neutrino experiments and accelerator and reactor experiments. The masses are tiny, fractions of an eV. The "see-saw mechanism" is a natural way to generate such masses. One supposes that in addition to the neutrinos of the Standard Model, there are some SU(2)xU(1)-singlet neutrinos, N. Nothing forbids these from obtaining a large mass. This could be of order MGUT, for example, or a bit smaller. These neutrinos could also couple to the left handed doublets νL, just like right handed charged leptons. Assuming that these couplings are not particularly small, one would obtain a mass matrix, in the {N, νL} basis, of the form

$$M_V = \begin{pmatrix} M & M_W \\ M_W^T & 0 \end{pmatrix}$$

This matrix has an eigenvalue $\dfrac{M_W^2}{M}$.

The latter number is of the order needed to explain the neutrino anomaly for M ~ 1013 or so, i.e. not wildly different than the GUT scale and other scales which have been proposed for new physics. For leptogenesis (Fukugita and Yanagida, 1986), what is important in this model is that the couplings of N break lepton number. N is a heavy particle; it can decay both to h + ν and h + ν-bar, for example. The partial widths to each of these final states need not be the same. CP violation can enter through phases in the Yukawa couplings and mass matrices of the N's.

As the universe cools through temperatures of order the of masses of the N's, they drop out of equilibrium, and their decays can lead to an excess of neutrinos over antineutrinos. Detailed predictions can be obtained by integrating a suitable set of Boltzmann equations. These decays produce a net lepton number, but not baryon number (and hence a net B – L). The resulting lepton number will be further processed by sphaleron interactions, yielding a net lepton and baryon number (recall that sphaleron interactions preserve B – L, but violate B and L separately). Reasonable values of the neutrino parameters give asymmetries of the order we seek to explain.

It is interesting to ask: assuming that these processes are the source of the observed asymmetry, how many parameters which enter into the computation can be measured, i.e. can we relate the observed number to microphysics. It is likely that, over time, many of the parameters of the light neutrino mass matrices, including possible CPviolating effects, will be measured. But while these measurements determine some of the couplings and masses, they are not, in general, enough. In order to give a precise calculation, analogous to the calculations of nucleosynthesis, of the baryon number density, one needs additional information about the masses of the fields N. One either

requires some other (currently unforseen) experimental access to this higher scale physics, or a compelling theory of neutrino mass in which symmetries, perhaps, reduce the number of parameters.

Production by Coherent Motion of Scalar Fields

The formation of an AD condensate can occur quite generically in cosmological models. Also, the AD scenario potentially can give rise simultaneously to the ordinary matter and the dark matter in the universe. This can explain why the amounts of luminous and dark matter are surprisingly close to each other, within one order of magnitude. If the two entities formed in completely unrelated processes (for example, the baryon asymmetry from leptogenesis, while the dark matter from freeze-out of neutralinos), the observed relation $\Omega_{DARK} \sim \Omega_{baryon}$ is fortuitous.

In supersymmetric theories, the ordinary quarks and leptons are accompanied by scalar fields. These scalar fields carry baryon and lepton number. A coherent field, i.e., a large classical value of such a field, can in principle carry a large amount of baryon number. As we will see, it is quite plausible that such fields were excited in the early universe. To understand the basics of the mechanism, consider first a model with a single complex scalar field. Take the Lagrangian to be

$$L = \left| \partial_\mu \phi \right|^2 - m^2 \left| \phi \right|^2$$

This Lagrangian has a symmetry, $\varphi \to e^{i\alpha\varphi}$, and a corresponding conserved current, which we will refer to as baryon current:

$$j_B^\mu = i \left(\phi^* \partial^\mu \phi - \phi \, \partial^\mu \phi^* \right).$$

It also possesses a "CP" symmetry: $\varphi \to \varphi*$. With supersymmetry in mind, we will think of m as of order M_W.

Let us add interactions in the following way, which will closely parallel what happens in the supersymmetric case. Include a set of quartic couplings:

$$L_I = \lambda \left| \phi \right|^4 + \in \phi^3 \phi^* + \delta \phi^4 + c.c.$$

These interactions clearly violate B. For general complex ε and δ, they also violate CP. In supersymmetric theories, as we will shortly see, the couplings will be extremely small. In order that these tiny couplings lead to an appreciable baryon number, it is necessary that the fields, at some stage, were very large.

To see how the cosmic evolution of this system can lead to a non-zero baryon number, first note that at very early times, when the Hubble constant, H $\gg$ m, the mass of the field is irrelevant. It is thus reasonable to suppose that at this early time $\varphi \to \varphi_o \gg 0$. How does the field then evolve? First ignore the quartic interactions. In the expanding universe, the equation of motion for the field is as usual

$$\ddot{\phi} + 3H\dot{\phi} + \frac{\partial V}{\partial \phi} = 0.$$

At very early times, H $\gg$ m, and so the system is highly overdamped and essentially frozen at φo. At this point, B = 0.

Once the universe has aged enough that H $\ll$ m, φ begins to oscillate. Substituting H = 1/2t or H = 2/3t for the radiation and matter dominated eras, respectively, one finds that

$$\phi = \begin{cases} \dfrac{\phi_o}{\left(mt \right)^{3/2}} \sin\left(mt \right) \ \left(\text{radiation} \right) \\[2ex] \dfrac{\phi_o}{\left(mt \right)} \sin\left(mt \right) \ \left(\text{matter} \right). \end{cases}$$

In either case, the energy behaves, in terms of the scale factor, R(t), as

$$E \approx m^2 \phi_O^2 \left(\frac{R_O}{R} \right)^3$$

Now let's consider the effects of the quartic couplings. Since the field amplitude damps with time, their significance will decrease with time. Suppose, initially, that $\varphi = \varphi_o$ is real. Then the imaginary part of φ satisfies, in the approximation that ε and δ are small,

$$\ddot{\phi}_i + 3H\dot{\phi}_t + m^2 \phi_t \approx \text{Im}(\in + \delta) \phi_r^3.$$

For large times, the right hand falls as $t^{-9/2}$, whereas the left hand side falls off only as $t^{-3/2}$. As a result, baryon number violation becomes negligible. The equation goes over to the free equation, with a solution of the form

$$\phi_i = a_r \frac{\text{Im}(\in + \delta) \phi_o^3}{m^2 (mt)^{3/4}} \sin(mt + \delta_r) \,(\text{radiation}), \quad \phi_i = a_m \frac{\text{Im}(\in + \delta) \phi_o^3}{m^3 t} \sin(mt + \delta_m) \,(\text{matter}),$$

The constants can be obtained numerically, and are of order unity

$$a_r = 0.85 \qquad a_m = 0.85 \qquad \delta_r = -0.91 \qquad \delta_m = 1.54.$$

But now we have a non-zero baryon number; substituting in the expression for the current,

$$n_B = 2a_r \, \text{Im}(\in + \delta) \frac{\phi_o^2}{m(mt)^2} \sin(\delta_r + \pi/8) \,(\text{radiation})$$

$$n_B = 2a_m \, \text{Im}(\in + \delta) \frac{\phi_o^2}{m(mt)^2} \sin(\delta_m) \,(\text{matter}).$$

Two features of these results should be noted. First, if ε and δ vanish, n_B vanishes. If they are real, and φ_o is real, n_B vanishes. It is remarkable that the Lagrangian parameters can be real, and yet φ_o can be complex, still giving rise to a net baryon number. Supersymmetry breaking in the early universe can naturally lead to a very large value for a scalar field carrying B or L. Finally, as expected, n_B is conserved at late times.

This mechanism for generating baryon number could be considered without supersymmetry. In that case, it begs several questions:

- What are the scalar fields carrying baryon number?

- Why are the φ^4 terms so small?

- How are the scalars in the condensate converted to more familiar particles?

In the context of supersymmetry, there is a natural answer to each of these questions. First, there are scalar fields (squarks and sleptons) carrying baryon and lepton number. Second, in the limit that supersymmetry is unbroken, there are typically directions in the field space in which the quartic terms in the potential vanish. Finally, the scalar quarks and leptons will be able to decay (in a baryon and lepton number conserving fashion) to ordinary quarks.

In addition to topologically stable solutions to the field equations such as strings or monopoles, it is sometimes also possible to find non-topological solutions, called **Q-balls**, which can form as part of the Affleck-Dine condensate. These are usually unstable and could decay to the dark matter, but in some theories they are stable and could be the dark matter. The various possibilities are summarized as follows:

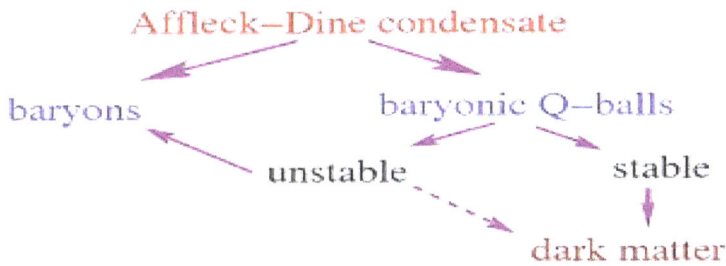

The parameter space of the MSSM consistent with LSP dark matter is very different, depending on whether the LSPs froze out of equilibrium or were produced from the evaporation of AD baryonic Q-balls. If supersymmetry is discovered, one will be able to determine the properties of the LSP experimentally. This will, in turn, provide some information on the how the dark-matter SUSY particles could be produced. The discovery of a Higgsino-like LSP would be a evidence in favor of Affleck–Dine baryogenesis. This is a way in which we might be able to establish the origin of matter-antimatter asymmetry.

References

- Vilenkin, Alex (2007). Many worlds in one : the search for other universes. New York: Hill and Wang, A division of Farrar, Straus and Giroux. p. 19. ISBN 978-0-8090-6722-0

- Percival, Will J.; et al. (March 2007). "The Shape of the Sloan Digital Sky Survey Data Release 5 Galaxy Power Spectrum". The Astrophysical Journal. 657 (2): 645–663. arXiv:astro-ph/0608636. Bibcode:2007ApJ...657..645P. doi:10.1086/510615

- Science 20 June 2003:Vol. 300. no. 5627, pp. 1914 - 1918Throwing Light on Dark Energy, Robert P. Kirshner. Retrieved December 2006

- Das, Saurya; Bhaduri, Rajat K (21 May 2015). "Dark matter and dark energy from a Bose–Einstein condensate". Classical and Quantum Gravity. 32 (10): 105003. arXiv:1411.0753. Bibcode:2015CQGra..32j5003D. doi:10.1088/0264-9381/32/10/105003

- Éduard Abramovich Tropp; Viktor Ya. Frenkel; Artur Davidovich Chernin (1993). Alexander A. Friedmann: The Man who Made the Universe Expand. Cambridge University Press. p. 219. ISBN 0-521-38470-2

- Planck Collaboration (2013). "Planck 2013 results. XVI. Cosmological parameters". Astronomy & Astrophysics. 571: A16. arXiv:1303.5076 [astro-ph.CO]. Bibcode:2014A&A...571A..16P. doi:10.1051/0004-6361/201321591

- "Simple but challenging: the Universe according to Planck". ESA Science & Technology. October 5, 2016 [March 21, 2013]. Retrieved October 29, 2016

- Perlmutter, S.; et al. (June 1999). "Measurements of Omega and Lambda from 42 High-Redshift Supernovae". The Astrophysical Journal. 517 (2): 565–586. arXiv:astro-ph/9812133. Bibcode:1999ApJ...517..565P. doi:10.1086/307221

Chapter 4

Cosmic Backgrounds

The cosmic background radiation is the electromagnetic radiation from the event of the big bang. The origin of the radiation is dependent on the spectrum region under observation. This chapter discusses in detail the background radiation of the universe such as the cosmic microwave background, cosmic infrared background and cosmic neutrino background.

Cosmic Microwave Background

The cosmic microwave background (or CMB) fills the entire Universe and is leftover radiation from the Big Bang. When the Universe was born, nearly 14 billion years ago, it was filled with hot plasma of particles (mostly protons, neutrons, and electrons) and photons (light). In particular, for roughly the first 380,000 years, the photons were constantly interacting with free electrons, meaning that they could not travel long distances. That means that the early Universe was opaque, like being in fog.

However, the Universe was expanding and as it expanded, it cooled, as the fixed amount of energy within it was able to spread out over larger volumes. After about 380,000 years, it had cooled to around 3000 Kelvin (approximately 2700°C) and at this point, electrons were able to combine with protons to form hydrogen atoms, and the temperature was too low to separate them again. In the absence of free electrons, the photons were able to move unhindered through the Universe: it became transparent.

Over the intervening billions of years, the Universe has expanded and cooled greatly. Due to the expansion of space, the wavelengths of the photons have grown (they have been 'redshifted') to roughly 1 millimetre and thus their effective temperature has decreased to just 2.7 Kelvin, or around -270°C, just above absolute zero. These photons fill the Universe today (there are roughly 400 in every cubic centimetre of space) and create a background glow that can be detected by far-infrared and radio telescopes.

Importance of Cosmic Microwave Background

The cosmic microwave background (CMB) is the furthest back in time we can explore using light. It formed about 380,000 years after the Big Bang and imprinted on it are traces of the seeds from which the stars and galaxies we can see today eventually formed. Hidden in the pattern of the radiation is a complex story that helps scientists to understand the history of the Universe both before and after the CMB was released.

Background of Cosmic Microwave Background

The existence of the cosmic microwave background (CMB) was postulated on theoretical grounds in the late 1940s by George Gamow, Ralph Alpher, and Robert Herman, who were studying the consequences of the nucleosynthesis of light elements, such as hydrogen, helium and lithium, at very early times in the Universe. They realised that, in order to synthesise the nuclei of these elements, the early Universe needed to be extremely hot and that the leftover radiation from this 'hot Big Bang' would permeate the Universe and be detectable even today as the CMB. Due to the expansion of the Universe, the temperature of this radiation has become lower and lower – they estimated at most 5 degrees above absolute zero (5 K), which corresponds to microwave wavelengths. It wasn't until 1964 that it was first detected – accidentally – by Arno Penzias and Robert Wilson, using a large radio antenna in New Jersey, a discovery for which they were awarded the Nobel Prize in Physics in 1978.

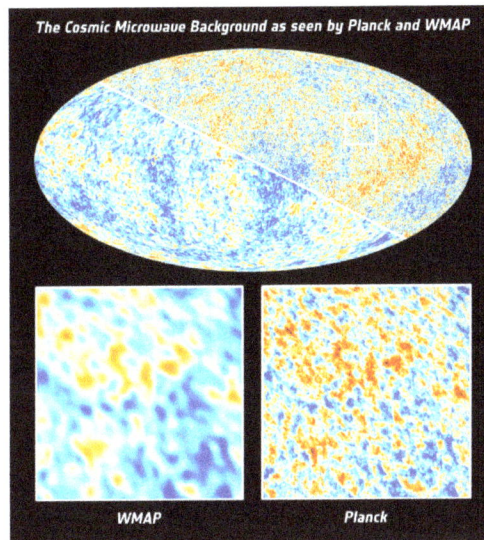

CMB as seen by Planck and WMAP

Space Missions that have Studied the Cosmic Microwave Background

The first space mission specifically designed to study the cosmic microwave background (CMB) was the Cosmic Background Explorer (COBE), launched by NASA in 1989. Among its key discoveries were that averaged across the whole sky, the CMB shows a spectrum that conforms extremely precisely to a so-called 'black body' (i.e. pure thermal radiation) at a temperature of 2.73 Kelvin, but that it also shows very small temperature fluctuations on the order of 1 part in 100,000 across the sky. These findings were rewarded with the award of the 2006 Nobel Prize in Physics to John Mather and George Smoot.

NASA's second generation space mission, the Wilkinson Microwave Anisotropy Probe (WMAP) was launched in 2001 to study these very small fluctuations in much more detail. The fluctuations were imprinted on the CMB at the moment where the photons and matter decoupled 380,000 years after the Big Bang, and reflect slightly higher and lower densities in the primordial Universe. These fluctuations were originated at an earlier epoch – immediately after the Big Bang – and would later grow, under the effect of gravity, giving rise to the large-scale structure (i.e. clusters and superclusters of galaxies) that we see around us today. WMAP's results have helped determine the proportions of the fundamental constituents of the Universe and to establish the standard model of cosmology prevalent today, and its scientists, headed by Charles Bennett, have garnered many prizes in physics in the intervening years.

Finally, ESA's Planck was launched in 2009 to study the CMB in even greater detail than ever before. It covers a wider frequency range in more bands and at higher sensitivity than WMAP, making it possible to make a much more accurate separation of all of the components of the submillimetre and microwave wavelength sky, including many foreground sources such as the emission from our own Milky Way Galaxy. This thorough picture thus reveals the CMB and its tiny fluctuations in much greater detail and precision than previously achieved. The aim of Planck is to use this greater sensitivity to prove the standard model of cosmology beyond doubt or, more enticingly, to search for deviations from the model which might reflect new physics beyond it.

Structure of Cosmic Microwave Background

The cosmic microwave background (CMB) is detected in all directions of the sky and appears to microwave telescopes as an almost uniform background. Planck's predecessors (NASA's COBE and WMAP missions) measured the temperature of the CMB to be 2.726 Kelvin (approximately -270 degrees Celsius) almost everywhere on the sky. The 'almost' is the most important factor here, because tiny fluctuations in the temperature, by just a fraction of a degree, represent differences in densities of structure, on both small and large scales, that were present right after the Universe formed. They can be imagined as seeds for where galaxies would eventually grow. Planck's instrument detectors are so sensitive that temperature variations of a few millionths of a degree are distinguishable, providing greater insight to the nature of the density fluctuations present soon after the birth of the Universe.

The Standard Model of Cosmology and its Relation to Cosmic Microwave Background

The standard model of cosmology rests on the assumption that, on very large scales, the Universe is homogeneous and isotropic, meaning that its properties are very similar at every point and that there are no preferential directions in space. In this model, the Universe was born nearly 14 billion years ago: at this time, its density and temperature were extremely high – a state referred to as 'hot Big Bang'. The Universe has been expanding ever since, as demonstrated by observations performed since the late 1920s. The rich variety of structure that we can observe on relatively small scales is the result of minuscule, random fluctuations that were embedded during cosmic inflation – an early period of accelerated expansion that took place immediately after the hot Big Bang – and that would later grow under the effect of gravity into galaxies and galaxy clusters.

The standard model of cosmology was derived from a number of different astronomical observations based on entirely different physical processes. To reconcile the data with theory, however, cosmologists have added two additional components that lack experimental confirmation: dark matter, an invisible matter component whose web-like distribution on large scales constitutes the scaffold where galaxies and other cosmic structure formed; and dark energy, a mysterious component that permeates the Universe and is driving its currently accelerated expansion. The standard model of cosmology can be described by a relatively small number of parameters, including: the density of ordinary matter, dark matter and dark energy, the speed of cosmic expansion at the present epoch (also known as the Hubble constant), the geometry of the Universe, and the relative amount of the primordial fluctuations embedded during inflation on different scales and their amplitude.

Different values of these parameters produce a different distribution of structures in the Universe, and a different corresponding pattern of fluctuations in the CMB. By looking at the CMB, Planck can help astronomers extract the parameters that describe the state of the Universe soon after it formed and how it evolved over billions of years.

Relationship to the Big Bang

The cosmic microwave background radiation and the cosmological redshift-distance relation are together regarded as the best available evidence for the Big Bang theory. Measurements of the CMB have made the inflationary Big Bang theory the Standard Cosmological Model. The discovery of the CMB in the mid-1960s curtailed interest in alternatives such as the steady state theory.

The CMB essentially confirms the Big Bang theory. In the late 1940s Alpher and Herman reasoned that if there was a big bang, the expansion of the universe would have stretched and cooled the high-energy radiation of the very early universe into the microwave region of the electromagnetic spectrum, and down to a temperature of about 5 K. They were slightly off with their estimate, but they had exactly the right idea. They predicted the CMB. It took another 15 years for Penzias and Wilson to stumble into discovering that the microwave background was actually there.

The CMB gives a snapshot of the universe when, according to standard cosmology, the temperature dropped enough to allow electrons and protons to form hydrogen atoms, thereby making the universe nearly transparent to radiation because light was no longer being scattered off free electrons. When it originated some 380,000 years after the Big Bang—this time is generally known as the "time of last scattering" or the period of recombination or decoupling—the temperature of the universe was about 3000 K. This corresponds to an energy of about 0.26 eV, which is much less than the 13.6 eV ionization energy of hydrogen.

Since decoupling, the temperature of the background radiation has dropped by a factor of roughly 1,100 due to the expansion of the universe. As the universe expands, the CMB photons are redshifted, causing them to decrease in energy. The temperature of this radiation stays inversely proportional to a parameter that describes the relative expansion of the universe over time, known as the scale length. The temperature T_r of the CMB as a function of redshift, z, can be shown to be proportional to the temperature of the CMB as observed in the present day (2.725 K or 0.2348 meV):

$$T_r = 2.725(1 + z)$$

Primary Anisotropy

The anisotropy, or directional dependency, of the cosmic microwave background is divided into two types: primary anisotropy, due to effects that occur at the last scattering surface and before; and secondary anisotropy, due to effects such as interactions of the background radiation with hot gas or gravitational potentials, which occur between the last scattering surface and the observer.

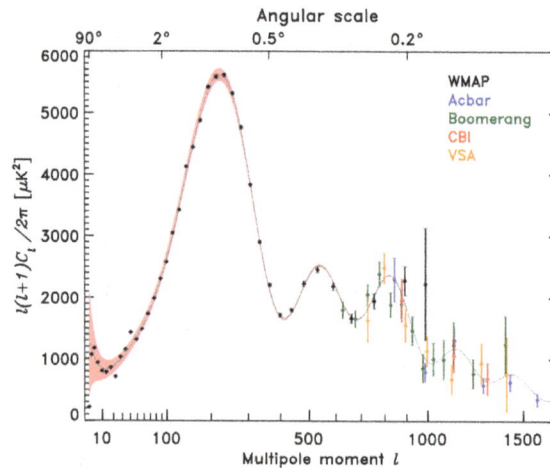

The power spectrum of the cosmic microwave background radiation temperature anisotropy in
terms of the angular scale (or multipole moment). The data shown comes from the WMAP,
Acbar Boomerang, CBI, and VSA instruments. Also shown is a theoretical model (solid line).

The structure of the cosmic microwave background anisotropies is principally determined by
two effects: acoustic oscillations and diffusion damping (also called collisionless damping or Silk
damping). The acoustic oscillations arise because of a conflict in the photon–baryon plasma in the
early universe. The pressure of the photons tends to erase anisotropies, whereas the gravitational
attraction of the baryons, moving at speeds much slower than light, makes them tend to collapse to
form overdensities. These two effects compete to create acoustic oscillations, which give the microwave background its characteristic peak structure. The peaks correspond, roughly, to resonances
in which the photons decouple when a particular mode is at its peak amplitude.

The peaks contain interesting physical signatures. The angular scale of the first peak determines
the curvature of the universe (but not the topology of the universe). The next peak—ratio of the odd
peaks to the even peaks—determines the reduced baryon density. The third peak can be used to get
information about the dark-matter density.

The locations of the peaks also give important information about the nature of the primordial
density perturbations. There are two fundamental types of density perturbations called *adiabatic*
and *isocurvature*. A general density perturbation is a mixture of both, and different theories that
purport to explain the primordial density perturbation spectrum predict different mixtures.

Adiabatic density perturbations

In an adiabatic density perturbation, the fractional additional number density of each type
of particle (baryons, photons ...) is the same. That is, if at one place there is a 1% higher
number density of baryons than average, then at that place there is also a 1% higher number density of photons (and a 1% higher number density in neutrinos) than average. Cosmic inflation predicts that the primordial perturbations are adiabatic.

Isocurvature density perturbations

In an isocurvature density perturbation, the sum (over different types of particle) of the
fractional additional densities is zero. That is, a perturbation where at some spot there is

1% more energy in baryons than average, 1% more energy in photons than average, and 2% *less* energy in neutrinos than average, would be a pure isocurvature perturbation. Cosmic strings would produce mostly isocurvature primordial perturbations.

The CMB spectrum can distinguish between these two because these two types of perturbations produce different peak locations. Isocurvature density perturbations produce a series of peaks whose angular scales (l values of the peaks) are roughly in the ratio 1:3:5:, while adiabatic density perturbations produce peaks whose locations are in the ratio 1:2:3:. Observations are consistent with the primordial density perturbations being entirely adiabatic, providing key support for inflation, and ruling out many models of structure formation involving, for example, cosmic strings.

Collisionless damping is caused by two effects, when the treatment of the primordial plasma as fluid begins to break down:

- the increasing mean free path of the photons as the primordial plasma becomes increasingly rarefied in an expanding universe,

- the finite depth of the last scattering surface (LSS), which causes the mean free path to increase rapidly during decoupling, even while some Compton scattering is still occurring.

These effects contribute about equally to the suppression of anisotropies at small scales and give rise to the characteristic exponential damping tail seen in the very small angular scale anisotropies.

The depth of the LSS refers to the fact that the decoupling of the photons and baryons does not happen instantaneously, but instead requires an appreciable fraction of the age of the universe up to that era. One method of quantifying how long this process took uses the *photon visibility function* (PVF). This function is defined so that, denoting the PVF by $P(t)$, the probability that a CMB photon last scattered between time t and $t + dt$ is given by $P(t)\,dt$.

The maximum of the PVF (the time when it is most likely that a given CMB photon last scattered) is known quite precisely. The first-year WMAP results put the time at which $P(t)$ has a maximum as 372,000 years. This is often taken as the "time" at which the CMB formed. However, to figure out how *long* it took the photons and baryons to decouple, we need a measure of the width of the PVF. The WMAP team finds that the PVF is greater than half of its maximal value (the "full width at half maximum", or FWHM) over an interval of 115,000 years. By this measure, decoupling took place over roughly 115,000 years, and when it was complete, the universe was roughly 487,000 years old.

Late Time Anisotropy

Since the CMB came into existence, it has apparently been modified by several subsequent physical processes, which are collectively referred to as late-time anisotropy, or secondary anisotropy. When the CMB photons became free to travel unimpeded, ordinary matter in the universe was mostly in the form of neutral hydrogen and helium atoms. However, observations of galaxies today seem to indicate that most of the volume of the intergalactic medium (IGM) consists of ionized material (since there are few absorption lines due to hydrogen atoms). This implies a period of reionization during which some of the material of the universe was broken into hydrogen ions.

The CMB photons are scattered by free charges such as electrons that are not bound in atoms. In an ionized universe, such charged particles have been liberated from neutral atoms by ionizing (ultraviolet) radiation. Today these free charges are at sufficiently low density in most of the volume of the universe that they do not measurably affect the CMB. However, if the IGM was ionized at very early times when the universe was still denser, then there are two main effects on the CMB:

1. Small scale anisotropies are erased. (Just as when looking at an object through fog, details of the object appear fuzzy.)

2. The physics of how photons are scattered by free electrons (Thomson scattering) induces polarization anisotropies on large angular scales. This broad angle polarization is correlated with the broad angle temperature perturbation.

Both of these effects have been observed by the WMAP spacecraft, providing evidence that the universe was ionized at very early times, at a redshift more than 17. The detailed provenance of this early ionizing radiation is still a matter of scientific debate. It may have included starlight from the very first population of stars (population III stars), supernovae when these first stars reached the end of their lives, or the ionizing radiation produced by the accretion disks of massive black holes.

The time following the emission of the cosmic microwave background—and before the observation of the first stars—is semi-humorously referred to by cosmologists as the dark age, and is a period which is under intense study by astronomers.

Two other effects which occurred between reionization and our observations of the cosmic microwave background, and which appear to cause anisotropies, are the Sunyaev–Zel'dovich effect, where a cloud of high-energy electrons scatters the radiation, transferring some of its energy to the CMB photons, and the Sachs–Wolfe effect, which causes photons from the Cosmic Microwave Background to be gravitationally redshifted or blueshifted due to changing gravitational fields.

Polarization

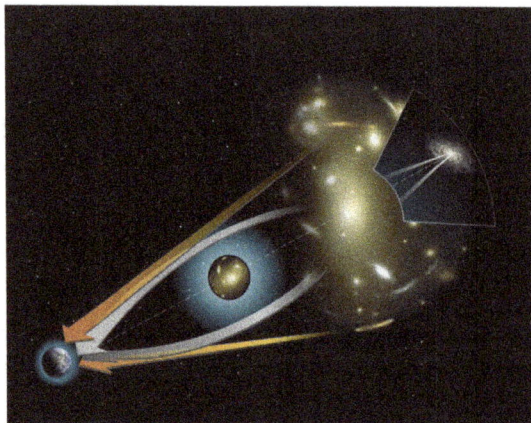

Artist's impression shows how light from the early universe is deflected by the gravitational lensing effect of massive cosmic structures forming B-modes as it travels across the universe.

The cosmic microwave background is polarized at the level of a few microkelvin. There are two types of polarization, called E-modes and B-modes. This is in analogy to electrostatics, in which the electric field (E-field) has a vanishing curl and the magnetic field (B-field) has a vanishing divergence. The E-modes arise naturally from Thomson scattering in a heterogeneous plasma. The B-modes are not produced by standard scalar type perturbations. Instead they can be created by two mechanisms: the first one is by gravitational lensing of E-modes, which has been measured by the South Pole Telescope in 2013; the second one is from gravitational waves arising from cosmic inflation. Detecting the B-modes is extremely difficult, particularly as the degree of foreground contamination is unknown, and the weak gravitational lensing signal mixes the relatively strong E-mode signal with the B-mode signal.

E-modes

E-modes were first seen in 2002 by the Degree Angular Scale Interferometer (DASI).

B-modes

Cosmologists predict two types of B-modes, the first generated during cosmic inflation shortly after the big bang, and the second generated by gravitational lensing at later times.

Primordial Gravitational Waves

Primordial gravitational waves are gravitational waves that could be observed in the polarisation of the cosmic microwave background and having their origin in the early universe. Models of cosmic inflation predict that such gravitational waves should appear; thus, their detection supports the theory of inflation, and their strength can confirm and exclude different models of inflation. It is the result of three things: inflationary expansion of space itself, reheating after inflation, and turbulent fluid mixing of matter and radiation.

On 17 March 2014 it was announced that the BICEP2 instrument had detected the first type of B-modes, consistent with inflation and gravitational waves in the early universe at the level of $r = 0.20+0.07 -0.05$, which is the amount of power present in gravitational waves compared to the amount of power present in other scalar density perturbations in the very early universe. Had this been confirmed it would have provided strong evidence of cosmic inflation and the Big Bang, but on 19 June 2014, considerably lowered confidence in confirming the findings was reported and on 19 September 2014 new results of the Planck experiment reported that the results of BICEP2 can be fully attributed to cosmic dust.

Gravitational Lensing

The second type of B-modes was discovered in 2013 using the South Pole Telescope with help from the Herschel Space Observatory. This discovery may help test theories on the origin of the universe. Scientists are using data from the Planck mission by the European Space Agency, to gain a better understanding of these waves.

In October 2014, a measurement of the B-mode polarization at 150 GHz was published by the POLARBEAR experiment. Compared to BICEP2, POLARBEAR focuses on a smaller patch of the sky and is less susceptible to dust effects. The team reported that POLARBEAR's measured B-mode polarization was of cosmological origin (and not just due to dust) at a 97.2% confidence level.

Microwave Background Observations

Subsequent to the discovery of the CMB, hundreds of cosmic microwave background experiments have been conducted to measure and characterize the signatures of the radiation. The most famous experiment is probably the NASA Cosmic Background Explorer (COBE) satellite that orbited in 1989–1996 and which detected and quantified the large scale anisotropies at the limit of its detection capabilities. Inspired by the initial COBE results of an extremely isotropic and homogeneous background, a series of ground- and balloon-based experiments quantified CMB anisotropies on smaller angular scales over the next decade. The primary goal of these experiments was to measure the angular scale of the first acoustic peak, for which COBE did not have sufficient resolution. These measurements were able to rule out cosmic strings as the leading theory of cosmic structure formation, and suggested cosmic inflation was the right theory. During the 1990s, the first peak was measured with increasing sensitivity and by 2000 the BOOMERanG experiment reported that the highest power fluctuations occur at scales of approximately one degree. Together with other cosmological data, these results implied that the geometry of the universe is flat. A number of ground-based interferometers provided measurements of the fluctuations with higher accuracy over the next three years, including the Very Small Array, Degree Angular Scale Interferometer (DASI), and the Cosmic Background Imager (CBI). DASI made the first detection of the polarization of the CMB and the CBI provided the first E-mode polarization spectrum with compelling evidence that it is out of phase with the T-mode spectrum.

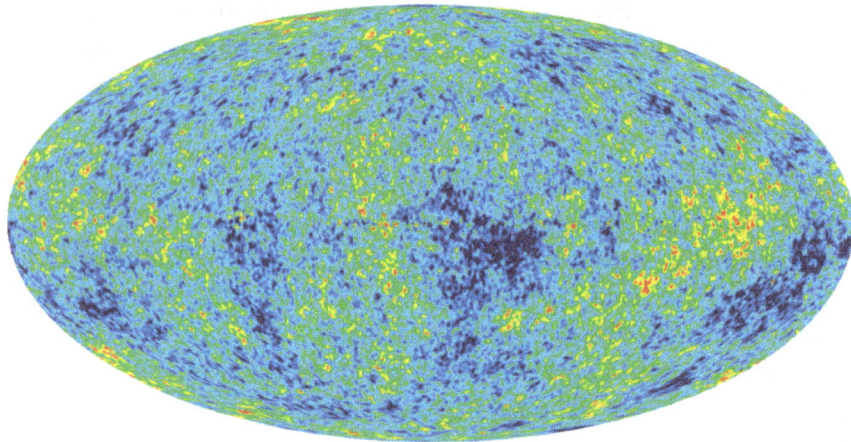

All-sky mollweide map of the CMB, created from 9 years of WMAP data

In June 2001, NASA launched a second CMB space mission, WMAP, to make much more precise measurements of the large scale anisotropies over the full sky. WMAP used symmetric, rapid-multi-modulated scanning, rapid switching radiometers to minimize non-sky signal noise. The first results from this mission, disclosed in 2003, were detailed measurements of the angular power spectrum at a scale of less than one degree, tightly constraining various cosmological parameters. The results are broadly consistent with those expected from cosmic inflation as well as various other competing theories, and are available in detail at NASA's data bank for Cosmic Microwave Background (CMB). Although WMAP provided very accurate measurements of the large scale angular fluctuations in the CMB (structures about as broad in the sky as the moon), it did not have the angular resolution to measure the smaller scale fluctuations which had been observed by former ground-based interferometers.

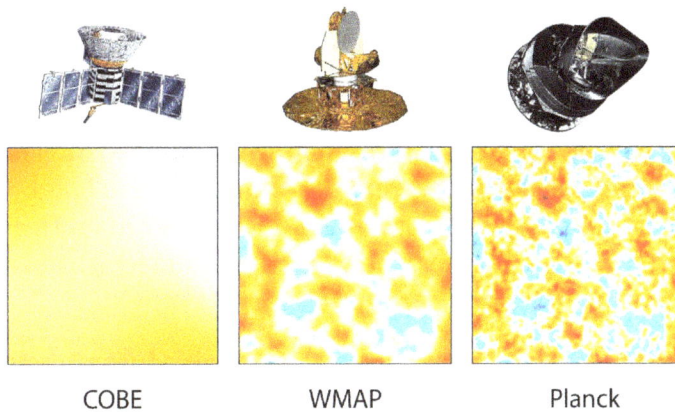

COBE   WMAP   Planck

Comparison of CMB results from COBE, WMAP and Planck

A third space mission, the ESA (European Space Agency) Planck Surveyor, was launched in May 2009 and performed an even more detailed investigation until it was shut down in October 2013. Planck employed both HEMT radiometers and bolometer technology and measured the CMB at a smaller scale than WMAP. Its detectors were trialled in the Antarctic Viper telescope as ACBAR (Arcminute Cosmology Bolometer Array Receiver) experiment—which has produced the most precise measurements at small angular scales to date—and in the Archeops balloon telescope.

On 21 March 2013, the European-led research team behind the Planck cosmology probe released the mission's all-sky map of the cosmic microwave background. The map suggests the universe is slightly older than researchers expected. According to the map, subtle fluctuations in temperature were imprinted on the deep sky when the cosmos was about 370000 years old. The imprint reflects ripples that arose as early, in the existence of the universe, as the first nonillionth of a second. Apparently, these ripples gave rise to the present vast cosmic web of galaxy clusters and dark matter. Based on the 2013 data, the universe contains 4.9% ordinary matter, 26.8% dark matter and 68.3% dark energy. On 5 February 2015, new data was released by the Planck mission, according to which the age of the universe is 13.799±0.021 billion years old and the Hubble constant was measured to be 67.74±0.46 (km/s)/Mpc.

Additional ground-based instruments such as the South Pole Telescope in Antarctica and the proposed Clover Project, Atacama Cosmology Telescope and the QUIET telescope in Chile will provide additional data not available from satellite observations, possibly including the B-mode polarization.

Data Reduction and Analysis

Raw CMBR data, even from space vehicles such as WMAP or Planck, contain foreground effects that completely obscure the fine-scale structure of the cosmic microwave background. The fine-scale structure is superimposed on the raw CMBR data but is too small to be seen at the scale of the raw data. The most prominent of the foreground effects is the dipole anisotropy caused by the Sun's motion relative to the CMBR background. The dipole anisotropy and others due to Earth's annual motion relative to the Sun and numerous microwave sources in the galactic plane and elsewhere must be subtracted out to reveal the extremely tiny variations characterizing the fine-scale structure of the CMBR background.

The detailed analysis of CMBR data to produce maps, an angular power spectrum, and ultimately cosmological parameters is a complicated, computationally difficult problem. Although computing a power spectrum from a map is in principle a simple Fourier transform, decomposing the map of the sky into spherical harmonics, in practice it is hard to take the effects of noise and foreground sources into account. In particular, these foregrounds are dominated by galactic emissions such as Bremsstrahlung, synchrotron, and dust that emit in the microwave band; in practice, the galaxy has to be removed, resulting in a CMB map that is not a full-sky map. In addition, point sources like galaxies and clusters represent another source of foreground which must be removed so as not to distort the short scale structure of the CMB power spectrum.

Constraints on many cosmological parameters can be obtained from their effects on the power spectrum, and results are often calculated using Markov Chain Monte Carlo sampling techniques.

CMBR Dipole Anisotropy

From the CMB data it is seen that the earth appears to be moving at 368±2 km/s relative to the reference frame of the CMB (also called the CMB rest frame, or the frame of reference in which there is no motion through the CMB). The Local Group (the galaxy group that includes the Milky Way galaxy) appears to be moving at 627±22 km/s in the direction of galactic longitude $l = 276°±3°$, $b = 30°±3°$. This motion results in an anisotropy of the data (CMB appearing slightly warmer in the direction of movement than in the opposite direction). From a theoretical point of view, the existence of a CMB rest frame breaks Lorentz invariance even in empty space far away from any galaxy. The standard interpretation of this temperature variation is a simple velocity red shift and blue shift due to motion relative to the CMB, but alternative cosmological models can explain some fraction of the observed dipole temperature distribution in the CMB.

Low Multipoles and other Anomalies

With the increasingly precise data provided by WMAP, there have been a number of claims that the CMB exhibits anomalies, such as very large scale anisotropies, anomalous alignments, and non-Gaussian distributions. The most longstanding of these is the low-l multipole controversy. Even in the COBE map, it was observed that the quadrupole ($l = 2$, spherical harmonic) has a low amplitude compared to the predictions of the Big Bang. In particular, the quadrupole and octupole ($l = 3$) modes appear to have an unexplained alignment with each other and with both the ecliptic plane and equinoxes. A number of groups have suggested that this could be the signature of new physics at the greatest observable scales; other groups suspect systematic errors in the data. Ultimately, due to the foregrounds and the cosmic variance problem, the greatest modes will never be as well measured as the small angular scale modes. The analyses were performed on two maps that have had the foregrounds removed as far as possible: the "internal linear combination" map of the WMAP collaboration and a similar map prepared by Max Tegmark and others. Later analyses have pointed out that these are the modes most susceptible to foreground contamination from synchrotron, dust, and Bremsstrahlung emission, and from experimental uncertainty in the monopole and dipole. A full Bayesian analysis of the WMAP power spectrum demonstrates that the quadrupole prediction of Lambda-CDM cosmology is consistent with the data at the 10% level and that the observed octupole is not remarkable. Carefully accounting for the procedure used to remove the foregrounds from the full sky map further reduces the significance of the alignment by ~5%. Recent observations with the Planck telescope, which is very much more sensitive than WMAP and has a larger angular resolution, record the same anomaly, and so instrumental error (but not foreground contamination) appears to be ruled out. Coincidence is a possible explanation, chief scientist from WMAP, Charles L. Bennett suggested coincidence and human psychology were involved, *"I do think there is a bit of a psychological effect; people want to find unusual things."*

Future Evolution

Assuming the universe keeps expanding and it does not suffer a Big Crunch, a Big Rip, or another similar fate, the cosmic microwave background will continue redshifting until it will no longer be detectable, and will be overtaken first by the one produced by starlight, and later by the background radiation fields of processes that are assumed will take place in the far future of the universe.

Cosmic Infrared Background

The cosmic infrared background (CIB) arises from accumulated emissions from early galaxy populations spanning a large range of redshifts. The earliest epoch for the production of this background occurred when star formation first began, and contributions to the CIB continued through the present epoch. The CIB is thus an integrated summary of the collective star forming events, star-burst activity and other luminous events in cosmic history to the present time. As photons move to observer they lose energy to cosmic expansion and any stellar emission from high redshift populations will now be seen in the infrared (mostly near-IR with $\lambda \lesssim$ a few µm , unless it comes from very cold stars at very high z). Emission from galactic dust will be shifted to still longer IR wavelengths. CIB thus probes the physics in he Universe between the present epoch and the last scattering surface and is complementary to its more famous cousin, the cosmic microwave background radiation (CMB) which probes mainly the physics at the last scattering.

Considerable effort has now gone into studying the luminosity sources during the most recent history of the universe (z < 2–3), but the period from recombination to the redshift of the Hubble Deep Field (HDF) remains largely an unexplored era because of the difficulty of detecting many distant galaxies over large areas of the sky. Significant progress in finding high redshift galaxies has been achieved using the Lyman dropout technique (Steidel et al 1996), but uncovering substantial populations of galaxies at   is extremely difficult and the total sample is still very small. It is there that the CIB measurements can provide critical information about the early history of the Universe largely inaccessible to telescopic studies. At more recent epochs galaxy evolution is also constrained by the measurements of the visible part of the extragalactic background light or EBL (Bernstein et al 2002a,b; cf. Mattila 2003).

Chronologically, the importance of the CIB and early predictions about its levels (Partridge & Peebles 1968) followed the discovery of the CMB which observationally established the Big Bang model for the origin and evolution of the Universe. Because of the Earth's atmosphere is was clear from the start that the measurement of the CIB must be done from space and even there the Solar system and Galactic foregrounds presented a formidable challenge. It took a while for technology to reach the required sensitivity and, when the first rocket and space based CIB measurements were conducted, the results were inconclusive (Matsumoto, Akiba & Murakami 1988, Noda et al 1992). Following its launch in 1983 the IRAS satellite was the first to conduct all-sky infrared measurements of point sources between 12 and 100 μm (Soifer, Neugebauer & Houck 1987). It revealed that galaxies were efficient infrared emitters, but its design was not optimized for diffuse background measurements. The COBE DIRBE instrument, launched in 1990, operated between 1.25 and 240 μm and was the first to be devoted specifically to CIB measurements (Hauser & Dwek 2001). It led to the first reliable measurements of and limits on the CIB over a wide range of infra-red bands and literally began the observational CIB era.

Observationally, the CIB is difficult to distinguish from the generally brighter foregrounds con-tributed by the local matter within the solar system, and the stars and ISM of the Galaxy. A num-ber of investigations have attempted to extract the isotropic component (mean level) of the CIB from ground- and satellite-based data as described below. This has in nearly all instances been a complicated task due to a lack of detailed knowledge of the absolute brightness levels and spatial variations across the sky for the many foregrounds that overlay the CIB signal. It is thus very im-portant to understand and estimate to high accuracy the various foreground emissions which need to subtracted or removed before uncovering the CIB.

The surface brightness of the CIB per unit wavelength will be denoted as I_λ, per unit frequency as I_V, and per logarithmic wavelength interval $F = \lambda I_\lambda = v I_{v,}$ and we call them all "flux." Through-out the paper $B_{v,}$ will denote the Planck black-body function per unit frequency.

Fluxes of astronomical sources are often measured in narrow band filters. The flux units common-ly used in astronomy for I_v are $Jy = 10^{-26}$ W/ m^2/ Hz.. The surface brightness of the CIB is usually given in units of either MJy/ sr or n W/ m^{-2} sr^{-1} The conversion between the two is:

$$1 \quad \frac{n\,W}{m^2\,sr} = \frac{3000}{\lambda(\mu m)}\frac{MJy}{sr}$$

The range of wavelengths used in this review is divided into the following groups: near-infrared (NIR) covers $1\ \mu m \lesssim \lambda \lesssim 5-10\,\mu m$. Mid-infrared (MIR) is defined to lie in $5-10\,\mu m \lesssim \lambda \lesssim 50-100\mu m$ range. We call the Far-infrared (FIR) the region corresponding to $50-100\mu m \lesssim \lambda \lesssim 500\mu m$, and beyond that will be sometimes referred to as sub-mm. These defi-nitions are neither exact nor unique and, although the band ranges cover (largely) different phys-ical processes, this division is used for convenience only.

CIB Anisotropies

Because of the difficulty of accurately accounting for the contributions of bright foregrounds such as Galactic stars, interplanetary and interstellar dust, which must be subtracted from the observed sky background (Arendt et al, 1998, Kelsall et al, 1998), Kashlinsky, Mather, Odenwald & Haus-er (1996) have proposed to measure the structure of the CIB or its fluctuations spectrum. For a relatively conservative set of assumptions about clustering of distant galaxies, fluctuations in the brightness of the CIB have a distinct spectral and spatial signal, and these signals can be more readily discerned than the actual mean level of the CIB. The most common source of luminosity in the universe arises in galaxies, whose clustering properties at the present times are fairly well known and are consistent with the $\wedge$CDM model predictions (Efstathiou, Sutherland & Maddox 1990, Percival et al 2002, Tegmark et al 2004). The CIB, being produced by clustered matter, must have fluctuations that reflect the clustered nature of the underlying sources of luminosity. This signature will have an angular correlation function (or angular power spectrum) that distinguishes it from local sources of background emission such as zodiacal light emission, and foreground stars in the Milky Way. Moreover, distant contributions of CIB will have a different redshift, and therefore spectral color, than nearby galaxies and sources of local emission. From galaxy evolution and cluster evolution models it is possible to match the predicted slopes for the power spectrum of the CIB against the power spectrum of the data.

On the largest angular scales (the dipole component), there would be an additional source of anisotropy due to our peculiar motion with respect to the inertial frame of the Universe. This component may be measurable over a certain range of wavelengths and is discussed after the more canonical source of CIB anisotropies, the galaxy clustering.

Anisotrophies form Galaxy Clurtering

Whenever CIB studies encompass relatively small parts of the sky (angular scales $\theta < 1$ sr) one can use Cartesian formulation of the Fourier analysis. The fluctuation in the CIB surface brightness can be defined as $\delta F(\theta) = F(\theta) - \langle F \rangle$, where $F = \lambda I_{\lambda}, \theta$ is the two dimensional coordinate on the sky and $\langle F \rangle$ is the ensemble average. The two-dimensional Fourier transform is $\delta F(\theta) = (2\pi)^{-2} \int \delta F_q \exp(-iq.\theta) d^2q$

If the fluctuation field, $\delta F(x)$ is a random variable, then it can be described by the moments of its probability distribution function. The first non-trivial moment is the projected 2-dimensional correlation function $C(\theta) = \langle \delta F(x+\theta) \delta F(x) \rangle$. The 2-dimensional power spectrum is $P_2(q) \equiv \langle |\delta F_q|^2 \rangle$, where the average is performed over all phases. The correlation function and the power spectrum are a pair of 2-dimensional Fourier transforms and for an isotropically distributed signal are related by

$$C(\theta) = \frac{1}{2\pi} \int_0^{\infty} P_2(q) J_0(q\theta) q dq,$$

$$P_2(q) = 2\pi \int_0^{\infty} C(\theta) J_0(q\theta) \theta d\theta,$$

where $J_n(x)$ is the n-th order cylindrical Bessel function. If the phases are random, then the distribution of the brightness is Gaussian and the correlation function (or its Fourier transform, the power spectrum) uniquely describes its statistics. In measurements with a finite beam, the intrinsic power spectrum is multiplied by the window function W of the instrument. Conversely, for the known beam window function, the power spectrum can be de-convolved by dividing the measured power spectrum by the beam window function.

Another useful and related quantity is the mean square fluctuation within a finite beam of angular radius ϑ, or zero-lag correlation signal, which is related to the power spectrum by

$$C(0) = \left\langle \left(\delta F_q\right)^2 \right\rangle \vartheta = \frac{1}{2\pi} \int_0^\infty P_2(q) W_{TH}(q\vartheta) q \, dq$$

$$\sim \frac{1}{2} q^2 P_2(q) \Big|_{q \sim \pi/2\vartheta}.$$

For a top-hat beam the window function is $W_{TH} = \left[2J(x)/x \right] = 0.5$ at x $\pi/2$ where $x = q\vartheta$, and hence the values of q^{-1} correspond to fluctuations on angular scales of diameter $\simeq \pi/q$.

At small angles < 1 sr, the CIB power spectrum is related to the CIB flux production rate, dF/dz, and the evolving 3-D power spectrum of galaxy clustering, $P_3(\kappa)$ via the Limber equation.

In the power spectrum formulation it can be written as (e.g. Kashlinsky & Odenwald 2000):

$$P_2(q) = \int \left(\frac{dF}{dz} \right)^2 \frac{1}{c \frac{dt}{dz} d_A^2(z)} P_3\left(qd_A^{-1}; z\right) dz$$

where $d_A(z)$ is the angular diameter distance and the integration is over the epoch of the sources-contributing to the CIB. Eq. 9 can be rewritten as:

$$P_2(q) = \frac{1}{c} \int \left(\frac{dI_{v'}}{dt} \right)^2 \frac{P_3\left(qd_A^{-1}; z\right)}{d_A^2} dt$$

This is equivalent to:

$$\frac{q^2 P_2(q)}{2\pi} = \pi t_0 \int \left(\frac{dI_{v'}}{dt} \right)^2 \Delta^2\left(qd_A^{-1}; z\right) dt$$

where t_0 is the time-length of the period over which the CIB is produced and

$$\Delta^2(\kappa) = \frac{1}{2\pi^2} \frac{\kappa^2 P_3(\kappa)}{ct_0}$$

is the fluctuation in number of sources within a volume $\kappa^{-2} ct_0.^3$

To within a factor of order unity, the square of the fractional fluctuation of the CIB on angular Scale $\simeq \pi/q$ is $\delta_{CIB}^2 = \left\langle \left(\delta I_v\right)^2 \right\rangle / I_v^2 \simeq I_v^{-2} q^2 P_2(q)/2\pi$. The meaning of eq. 11 becomes obvious if we assume $dI_{v'}/dt = $ constant during the lifetime of the emitters t_0. In this case the fractional fluctuation due to clustering of early galaxies becomes:

$$\delta_{\text{CIB}}^2 = \frac{\pi}{t_0} \int \Delta^2\left(q d_A^{-1}; z\right) dt$$

In other words, the fractional fluctuation on angular scale π/q in the CIB is given by the average value of the r.m.s. fluctuation from spatial clustering over a cylinder of length ct_0 and diameter.

The Cartesian formulation is equivalent to the spherical sky representation used in the cosmic microwave background (CMB) studies on small scales. It was used in some CIB analyses. In that case one expands the flux into spherical harmonics:

$$F(\theta,\phi) = \nu I_\nu = \sum_{l=0}^{\infty} \sum_{m=-l}^{l} a_{lm} Y_{lm}(\theta,\phi).$$ The correlation function of the

CIB, $C(\theta) = \langle \delta F(x). \delta F(x+\theta)\rangle$, is then $C(\theta) = \sum \frac{(2l+1)}{4\pi} C_l P_l(\cos\theta)$ with $C_l = \langle |a_l|^2\rangle \equiv (2l+1)^{-1} |a_{lm}|^2$

and P_l denoting the Legendre polynomials (e.g. Peebles 1980). The angular power spectrum $P_2(q)$, is the two-dimensional Fourier transform of $C(\theta)$ and for $l \gg 1$ is related

to the multipoles via $C_l = P_2\left(l + \frac{1}{2}\right)$. This follows because at $l \gg 1$ the Legendre polynomials can be approximated as Bessel functions, $P_l(\cos\theta) \simeq J_0\left((l+1/2)\theta\right)$. The magnitude

of the CIB fluctuation on scale π/l radian for large l is then $\sim \sqrt{l^2 C_l/2\pi}$. In the limit of small angles $(l \gg 1)$ the values of C_l's are related to the power spectrum of galaxy clustering and the CIB flux production rate via:

$$C_l = \frac{1}{c} \int \left(\frac{dI\nu'}{dt}\right)^2 \frac{1}{d_A^2(z)} P_3\left(\frac{l+\frac{1}{2}}{d_A(z)}; z\right) dt$$

In surveys with arcsec angular resolution it is possible to identify and remove galaxies brighter than some limiting magnitude, $m_{\text{lim.}}$ Because on average fainter galaxies are at higher z, by improving sensitivity and angular resolution, one can isolate contributions to the CIB fluctuations from progressively earlier epochs (Kashlinsky et al 2002). The luminosity density, $L = \int_0^{L(m_{\text{lim.}})} \Phi(L) dL$, is then peaked at some particular redshift - at lower the (brighter than apparent magnitude $m_{\text{lim.}}$) galaxies are removed and at larger redshifts the luminosity density gets dominated by the bright end of the luminosity function with a sharp drop-off in the galaxy number density. The following toy model is useful in estimating the effect: in the visible to near-IR bands the present day galaxy luminosity function is of the Schechter (1976) form $\Phi_* = \Phi_* L_*^{-1}(L/L_*)^{-\alpha} \exp(L/L_*)$. Measurements of the galaxy luminosity function from B to K bands indicate that within the statistical uncertainties $\alpha \simeq 1$ (Loveday et al 1992, Gardner et al 1997), leading to $L = \Phi_* L_* \left(1 - \exp\left[-L(m_{\text{lim.}})/L_*\right]\right)$. This then defines the redshift window in eqs. 10,4 which contributes most to the power spectrum of the CIB. Contribution from low z galaxies, for which $L(m_{\text{lim.}}) < L_*$ is $L \simeq \Phi_* L(m_{\text{lim.}}) \propto d_L^2(z)$. In practice the typical redshift at which most of the contribution arises can be estimated as $L(m_{\text{lim.}}) \sim L_*$.

If one can further remove galaxies lying in narrow bins, Δ_m around progressively fainter apparent magnitude m_{lim}, one can hope to isolate contributions to the CIB by epoch (Kashlinsky 1992, Kashlinsky et al 2002).

Shot Noise Fluctuations from Individual Galaxies

In addition to fluctuations from galaxy clustering there would also be a shot-noise component arising from discrete galaxies occasionally entering the beam. The relative amplitude of these shot-noise fluctuations will be $\sim N_{\text{beam}}^{-1/2}$ where N_{beam} is the average number of galaxies in the beam. This component is important in surveys with good angular resolution where $N_{\text{beam}} \lesssim$ a few.

If galaxies are removed down to some magnitude m, the shot noise contribution from the remaining sources to the flux variance, C(0), in measurements with beam of the area ω_{beam} steradian is given by:

$$\sigma_{\text{sn}}^2 = \frac{1}{\omega_{\text{beam}}} \int_m^\infty F^2(m) \frac{dN_{\text{gal}}}{dm} dm$$

Here $F(m) = F_0 10^{-0.4m}$ is the flux from galaxy of magnitude m and dN_{gal}/dm is the number of galaxies per steradian in the magnitude bin dm.

Fourier amplitudes of the shot noise are scale-independent and equation 8 implies that the shot noise contribution to the power spectrum of the CIB would be given by:

$$P_{\text{sn}} = \int_m^\infty F^2(m) \frac{dN_{\text{gal}}}{dm} dm$$

Because galaxy clustering has power spectrum that increases towards large scales, the shot noise component becomes progressively more important at smaller angular scales.

Cosmic Variance for CIB Aninotropies

Any measurement of the angular power spectrum will be affected by the sample or cosmic variance in much the same way as the cosmic microwave background measurement (Abbot & Wise 1984). This results from the fact that in the best of situations we can only observe 4π steradian leading to poor sampling of the long wavelength modes. If the power spectrum is determined from fraction f_{sky} of the sky by sampling in concentric rings of width Δ_q in angular wavenumber space, the relative uncertainty on $C_l = P_2(q)$ will be $N_q^{-1/2}$, where $N_q \propto q\Delta q$ is the number of ring elements in $[q; q+\Delta q]$ interval. Therefore, the relative uncertainty from cosmic variance in the measured power spectrum on scale $\theta \approx \pi/\theta$ will be:

$$\frac{\sigma_{P_2}^{CV}}{P_2}\bigg|\text{cosmic variance} \simeq \frac{1}{2\pi} \frac{\theta}{180^0} \sqrt{\frac{q}{\Delta q}} f_{\text{sky}}^{-1/2}$$

In order to get reliable and independent measurements at a given scale, it is useful to have narrow band $\Delta q/q \sim 0.05 - 0.1$. Therefore, for reliable measurements on scales up to θ, one has to cover an area a few times larger.

CIB Dipole Component

The dipole anisotropy of the CIB arises from our local motion with respect to the inertial frame of the Universe, rather than galaxy clustering. It carries important cosmological information and its amplitude and wavelength dependence can be predicted in a model-independent way and may one day be measurable.

If all of the dipole anisotropy of the CMB is produced by peculiar motion of the Sun and the Local Group with respect to the inertial frame provided by a distant observer (the last scattering surface in the case of the CMB or early epochs, high z, in the case of the CIB), the CIB should have dipole anisotropy of the corresponding amplitude and in the same direction. The amplitude of the dipole anisotropy can be characterized by the first term, C_1 in expanding the sky in spherical harmonics. Since I_ν/ν^3 is an optical constant along the ray's trajectory, the motion of the terrestrial observer at speed v_{pec} with respect to the observed background will produce dipole fluctuation of the amplitude $\delta I_\nu/\langle I_\nu\rangle = (3-\alpha_\nu)(v_{pec}/c)\cos\theta = C_1\cos\theta$ (e.g. Peebles & Wilkinson 1968). Here θ is the angle between the line-of-sight and the direction of motion and $\alpha_\nu \equiv \partial\ln I_\nu/\partial\ln\nu$ is the spectral index of the radiation.

For CMB measurements in the Rayleigh-Jeans part of the CMB spectrum the index is $\alpha_{CMB} \simeq 2$. Hence, the CIB dipole is related to that of the CMB via $C_{1,\,CIB} = (3-\alpha_\nu)C_{1,\,CMB}/T_{CMB}F_{CIB}$. The CMB dipole is known very accurately to be $C_{1,\,CIB} = 3.346 \pm 0.017$ mK (Bennett et al 2003). Hence the CIB dipole amplitude is expected to be:

$$\delta F_{CIB,\,dipole} \simeq 1.2\times10^{-3}\left(3-\alpha_\nu\right)F_{CIB}$$

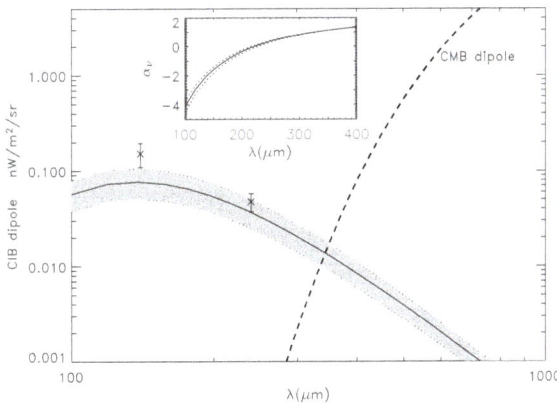

Figure: The insert shows the mean spectral index of the far-IR CIB according to eq. 24 (solid line) with its approximate uncertainty (dotted lines). Figure shows the expected amplitude of the CIB dipole anisotropy normalized to the CMB dipole.

uncertainty. Crosses with errors show the dipole expected for the DIRBE measured levels of the CIB at 140 and 240 μm. Thick dashed line shows the latest measurements of the CMB dipole.

In the far-IR they show that the energy spectrum of the CIB has a 'window' where α_ν is strongly negative figure above shows the spectral index of the CIB derived from the FIRAS based measurements of the CIB, and the resultant CIB dipole amplitude normalized to the observed CMB dipole. This component may be measurable at wavelengths where the CIB has a strongly negative α_ν and a sufficiently high flux level. The measurements show that and a sufficiently high flux level. The measurements show that α_ν is strongly negative for $\lambda < 200\,\mu m$ reaching the levels of $\alpha_\nu \sim -2$ to -4 there. Around $\sim 100\,\mu m$ the CIB spectrum has $\alpha_\nu \simeq -4$ and the CIB dipole should be $\simeq 7$ times more sensitive than the CMB dipole. The CIB flux at these wavelengths is $\sim 10-20\,n\,Wm^{-2}\,sr^{-1}$ and the CIB dipole at this range of wavelengths should have a non-negligible amplitude of $\sim 0.1-0.2\,n\,Wm^{-2}\,sr^{-1}$ and may be detectable in some future measurements. At longer wavelengths the CIB dipole will be difficult to measure due to steep increase in the residual uncertainty from the CMB dipole which is also shown in the figure.

The CIB dipole anisotropy could be additionally enhanced because the over-density that provides our peculiar acceleration also presumably has excess IR luminosity. Indeed, there are persistent claims of large bulks flows on scales of $\sim 150-200h^{-1}$ Mpc, whose direction roughly coincides with that of the CMB dipole. Furthermore, the dipole of the distribution of rich (Abell) clusters does not converge out to $\sim 200h^{-1}$ Mpc, while its direction roughly coincides with the CMB dipole (Scaramella et al. 1991).

The measurement of the CIB dipole anisotropy should be possible in the wavelength range 100–300 μm and will be important to provide additional information on the peculiar motions in the local part of the Universe and will serve as additional, and perhaps ultimate, test of the cosmological nature of any CIB detection at that wavelength.

Observation of the Cosmic Infrared Background

The detection of the CIB is both observationally and astrophysically very challenging. It has a very few characteristics which can be used to separate it from the foregrounds. One major point is, that the CIB must be isotropic, i.e. one has to measure the same CIB value all over the sky. It also lacks suspicious spectral features, since the final shape of its spectrum is the sum of the spectra of sources in the line of sight at various redshifts.

Direct Detection

Direct measurements are simple, but very difficult. One just has to measure the total incoming power, and determine the contribution of each sky background component. The measurement has to be repeated in many directions to determine the contribution of the foregrounds. After the removal of all other components the remaining power - if it is the same constant value in any direction - is the CIB at that specific wavelength. In practice, one needs an instrument which is able to perform absolute photometry, i.e. it has some mechanism to fully block incoming light for an accurate zero level determination (cold shutter). Since the instrument parts, including the shutter, have non-zero temperatures and emit in the infrared, this is a very difficult task.

The first, and still the most extensive, direct CIB measurements were performed by the DIRBE instrument of the COBE satellite. After the removal of the precisely determined zodiacal emission contribution (which was based on the measured annual variation) the remaining power at longer infrared wavelength contained basically two components: the CIB and the Galactic cirrus emission. The infrared surface brightness of the Galactic cirrus must correlate with the neutral hydrogen column densities, since they originate from the same, low density structure. After the removal of the HI-correlated part, the remaining surface brightness was identified as the cosmic infrared background at 60, 100, 140 and 240µm. At shorter wavelengths the CIB level could not be correctly determined.

Later, short wavelength DIRBE measurements at 2.2 and 3.5µ were combined with the Two Micron Sky Survey (2MASS) source count data, and this led to the detection of the CIB at these two wavelengths.

Fluctuation Studies

Since the CIB is an accumulated light of individual sources there is always a somewhat different number of sources in different directions in the field of view of the observer. This cause a variation (fluctuation) in the total amount of observed incoming flux among the different line of sights. These fluctuations are traditionally described by the two dimensional autocorrelation function, or by the corresponding Fourier power spectrum. The detection of fluctuations is easier than the direct CIB measurements, since one does not need to determine the absolute photometric zero point - fluctuations can be derived from differential measurements. On the other hand, fluctuations do not provide an immediate information on the CIB brightness. The measured fluctuation amplitudes either has to be confronted with a CIB model that has a prediction for the fluctuation / absolute level ratio, or it has to be compared with integrated differential light levels of source counts at the same wavelength.

The power spectrum of the CIB is usually presented in a spatial frequency [arcmin^{-1}] vs. fluctuation power [Jy^2 sr^{-1}] diagram. It is contaminated by the presence of the power spectrum of foreground components, so that the total power spectrum is:

$$P(f) = \Phi(f) \times [P_{CIB}(f) + P_{cirr}(f) + P_{ze}(f) + P_n(f)]$$

where $P(f)$, $P_{CIB}(f)$, P_{cirr}, $P_{ze}(f)$ and $P_n(f)$ are the total, CIB, Galactic cirrus, zodiacal emission and noise (instrument noise) power spectrum components, respectively, and Φ is the power spectrum of the telescope's point spread function.

For most of the infrared zodiacal emission fluctuation are negligible in the "cosmic windows", far from the ecliptic plane.

In the far-infrared the CIB power spectrum can be effectively used to separate it from its strongest foreground, the Galactic cirrus emission. The cirrus emission has a characteristic power spectrum of a power-law (that of a fractal spatial structure) $P(f) = P_o(f/f_o)^\alpha$, where P is the fluctuation power at the spatial frequency f, P_o is the fluctuation power at the reference spatial frequency f_o, and α is the spectral index. α was found to be $\alpha \approx -3$, which is much steeper than the power spectrum of the CIB at low spatial frequencies. The cirrus component can be identified in the power spectrum at low spatial frequencies and then removed from the whole spatial frequency range. The remaining power spectrum - after a careful correction for instrument effects - should be that of the CIB.

Autocorrelation and power spectrum studies resulted in the CIB fluctuation amplitudes at 1.25, 2.2, 3.5, 12-100µm based on the COBE/DIRBE measurements, and later at 90 and 170µm, based on the observations of the ISOPHOT instrument of the Infrared Space Observatory. Recently, the clustering of the galaxies have also been identified in the power spectrum at 160µm using this method.

Source Counts

Source counts gives the most extensive picture about the sources building up the CIB. In a source count one tries to detect as many point/compact sources in a certain field of view as possible: this is usually done at multiple wavelengths and is often complemented by other data, e.g. photometry at visual or sub-millimeter wavelengths. In this way, one has information on the broad band spectral characteristics of the detected sources, too. The detected point sources have to be distinguished from other contaminating sources, e.g. minor bodies in the Solar System, Galactic stars and cirrus knots (local density enhancements in the Galactic cirrus emission).

Source counts were important tasks for the recent infrared missions like 2MASS or the Infrared Space Observatory (ISO), and is still one of the most important questions the current and near future infrared space instruments (the Spitzer Space Telescope and the Herschel Space Observatory). While ISO was able to resolve about 3-10% of the total CIB light into individual sources (depending on the wavelength), Spitzer measurements have already detected ~30% of the CIB as sources, and this ratio is expected to be ~90% at some wavelengths with the Herschel Space Observatory.

Source count results support the "fast evolution" galaxy models. In these models galaxies nowadays look significantly different than they were at z=1...2, when they were coming through an intense star-formation phase. The source count results exclude the "steady state" scenarios, where z=1...2 galaxies look similar to those we see today in our cosmic neighborhood

Cosmic Neutrino Background

Cosmic neutrino background, low-energy neutrinos that pervade the universe. When the universe was one second old, it had cooled enough that neutrinos no longer interacted with ordinary matter. These neutrinos now form the cosmic neutrino background.

Cosmic neutrinos can be subdivided into three categories according to their origin and energy. These are the cosmic neutrino background (CNB, or CnB), stellar neutrinos, and cosmogenic neutrinos. Similar to the cosmic microwave background (CMB), induced at around 380,000 years after Big Bang, the CnB was induced at around 2 seconds after the Big Bang when the rate of their interactions with other particles became slower than the rate of expansion of the universe at that time. It is estimated that by now CnB has been cooled to ~1.9 °K temperature, or about 10^{-4} eV in energy, which makes them non-relativistic. The CnB has never been detected.

All stars shine as a result of nuclear fusion reaction in the stellar core. One necessary product of nuclear fusion is electron-neutrinos. Being weakly interacting, these neutrinos escape from the star. For example, the neutrino flux from the Sun at the Earth is about 10^{11} particles/cm²/sec. Solar neutrinos were first detected by Raymond Davis, who discovered neutrino flavor oscillations at the same time. Also when a star uses up its nuclear fuel and explodes as a supernova, it releases a huge number of neutrinos; about 99% of the exploding energy is carried by neutrinos. Masatoshi Koshiba led the effort in the first detection of supernova neutrinos of SN1987A in 1987. These marked the beginning of neutrino astronomy. Davis and Koshiba were jointly awarded the 2002 Nobel Prize in Physics for their contributions.

Finally, there also exist the extremely high-energy cosmogenic neutrinos. These are the neutrinos produced by the collisions between ultra high energy cosmic rays (UHECR) and CMB photons, also referred to as Greisen-Zatsepin-Kuzmin (GZK) neutrinos. The fact that UHECRs (mostly protons) with energies up to 10^{20} eV and CMB have been observed on Earth implies that GZK neutrinos with energies around 10^{17}-10^{19} eV must exist with a sufficient flux based on known, standard model particle physics. However, so far these neutrinos have not been observed yet. The world's largest cosmic neutrino observatory IceCube at South Pole have recently observed a pair of cosmic neutrinos at energies around 10^{15} eV, which were nicknamed by the IceCube Collaboration as "Bert" and "Ernie", after characters from the Sesame Street TV show. Later in 2013 an even more energetic neutrino was found, and it was nicknamed "Big Bird".

Derivation of the CvB Temperature

Given the temperature of the CMB, the temperature of the CvB can be estimated. Before neutrinos decoupled from the rest of matter, the universe primarily consisted of neutrinos, electrons, positrons, and photons, all in thermal equilibrium with each other. Once the temperature dropped to approximately 2.5 MeV, the neutrinos decoupled from the rest of matter. Despite this decoupling, neutrinos and photons remained at the same temperature as the universe expanded. However, when the temperature dropped below the mass of the electron, most electrons and positrons annihilated, transferring their heat and entropy to photons, and thus increasing the temperature of the photons. So the ratio of the temperature of the photons before and after the electron-positron annihilation is the same as the ratio of the temperature of the neutrinos and the photons today. To find this ratio, we assume that the entropy of the universe was approximately conserved by the electron-positron annihilation. Then using

$$\sigma \propto g T^3,$$

where σ is the entropy, g is the effective degrees of freedom and T is the temperature, we find that

$$\left(\frac{g_0}{g_1} \right)^{1/3} = \frac{T_1}{T_0},$$

where T_0 denotes the temperature before the electron-positron annihilation and T_1 denotes after. The factor g_0 is determined by the particle species:

- 2 for photons, since they are massless bosons

- $2 \times (7/8)$ each for electrons and positrons, since they are fermions.

g_1 is just 2 for photons. So

$$\frac{T_\nu}{T_\gamma} = \left(\frac{2}{2 + 2 \times 7/8 + 2 \times 7/8} \right)^{1/3} = \left(\frac{4}{11} \right)^{1/3}.$$

Given the current value of T_γ = 2.725 K, it follows that $T_\nu \approx 1.95$ K.

The above discussion is valid for massless neutrinos, which are always relativistic. For neutrinos with a non-zero rest mass, the description in terms of a temperature is no longer appropriate after they become non-relativistic; i.e., when their thermal energy $3/2\ kT_v$ falls below the rest mass energy $m_v c^2$. Instead, in this case one should rather track their energy density, which remains well-defined.

Indirect Evidence for the CvB

Relativistic neutrinos contribute to the radiation energy density of the universe ρ_R, typically parameterized in terms of the effective number of neutrino species N_v:

$$\rho_R = \frac{\pi^2}{15}T_\gamma^4(1+z)^4\left[1+\frac{7}{8}N_v\left(\frac{4}{11}\right)^{4/3}\right],$$

where z denotes the redshift. The first term in the square brackets is due to the CMB, the second comes from the CvB. The Standard Model with its three neutrino species predicts a value of $N_v \simeq 3.046$, including a small correction caused by a non-thermal distortion of the spectra during e^+-e^--annihilation. The radiation density had a major impact on various physical processes in the early universe, leaving potentially detectable imprints on measurable quantities, thus allowing us to infer the value of N_v from observations.

Big Bang Nucleosynthesis

Due to its effect on the expansion rate of the universe during Big Bang nucleosynthesis (BBN), the theoretical expectations for the primordial abundances of light elements depend on N_v. Astrophysical measurements of the primordial ^{4}He and ^{2}D abundances lead to a value of $N_v = 3.14+0.70-0.65$ at 68% c.l., in very good agreement with the Standard Model expectation.

CMB Anisotropies and Structure Formation

The presence of the CvB affects the evolution of CMB anisotropies as well as the growth of matter perturbations in two ways: due to its contribution to the radiation density of the universe (which determines for instance the time of matter-radiation equality), and due to the neutrinos' anisotropic stress which dampens the acoustic oscillations of the spectra. Additionally, free-streaming massive neutrinos suppress the growth of structure on small scales. The WMAP spacecraft's five-year data combined with type Ia supernova data and information about the baryon acoustic oscillation scale yielded $N_v = 4.34+0.88-0.86$ at 68% c.l., providing an independent confirmation of the BBN constraints. The Planck spacecraft collaboration has published the tightest bound to date on the effective number of neutrino species, at $N_v = 3.15\pm0.23$.

Indirect Evidence from Phase Changes to the Cosmic Microwave Background (CMB)

Big Bang cosmology makes many predictions about the CNB, and there is very strong indirect evidence that the cosmic neutrino background exists, both from Big Bang nucleosynthesis predictions of the helium abundance, and from anisotropies in the cosmic microwave background. One of these predictions is that neutrinos will have left a subtle imprint on the cosmic microwave background (CMB). It is well known that the CMB has irregularities. Some of the CMB fluctuations were roughly regularly spaced, because of the effect of baryon acoustic oscillation. In theory, the decoupled neutrinos should have had a very slight effect on the phase of the various CMB fluctuations.

In 2015, it was reported that such shifts had been detected in the CMB. Moreover the fluctuations corresponded to neutrinos of almost exactly the temperature predicted by Big Bang theory (1.96 +/-0.02K compared to a prediction of 1.95K), and exactly three types of neutrino, the same number of neutrino flavours currently predicted by the Standard Model.

Prospects for the Direct Detection of the CvB

Confirmation of the existence of these relic neutrinos may only be possible by directly detecting them using experiments on Earth. This will be difficult as the neutrinos which make up the CvB are non-relativistic, in addition to interacting only weakly with normal matter, and so any effect they have in a detector will be hard to identify. One proposed method of direct detection of the CvB is to use capture of cosmic relic neutrinos on tritium i.e. ^3H, leading to an induced form of beta decay. The neutrinos of the CvB would lead to the production of electrons via the reaction $v + {}^3\text{H} \rightarrow {}^3\text{He} + e^-$, while the main background comes from electrons produced via natural beta decay $^3\text{H} \rightarrow {}^3\text{He} + e^- + \bar{v}$. These electrons would be detected by the experimental apparatus in order to measure the size of the CvB. The latter source of electrons is far more numerous, however their maximum energy is smaller than the average energy of the CvB-electrons by twice the average neutrino mass. Since this mass is tiny, of the order of a few eVs or less, such a detector must have an excellent energy resolution in order to separate the signal from the background. One such proposed experiment is called PTOLEMY, which will be made up of 100g of tritium target. As of 2016 PTOLEMY prototype is being built.

The KATRIN Detection of the Cosmic Neutrino Background (CNB)

Fermi's Golden Rule gives for the Tritium beta-decay probability:

$$\Gamma^\beta_{\text{decay}}(^3\text{H}) =$$

$$\frac{1}{2\pi^3} \sum \int \left| <^3\text{He} | \text{T} | ^3\text{H} > \right|^2 2\pi\delta(E_v + E_e + E_f - E_i) \frac{d\vec{p}_e}{2\pi^3} \frac{d\vec{p}_v}{2\pi^3}$$

Theory : $T^\beta_{1/2} = \dfrac{\ln 2}{\Gamma^\beta_{\text{decay}}} = 12.32\,\text{yrs}$; Experiment : $T^\beta_{1/2} = 12.33\,\text{yrs}$.

The β-decay spectrum of the electron from the Tritium β-decay has the

form:

$$\frac{dN_e}{dE} = K\, F(E,Z)\, p_e E_e (E_o - E_e) \times$$

$$\sum_{j=1}^{3}\left|U_{ej}\right|^{2}\sqrt{\left(E_{0}-E_{e}\right)^{2}-m_{vj}^{2}}\,\theta\left(E_{0}-E_{e}-mvj\right)$$

with:

$$K=\text{const};\quad Q=18.562\text{ keV};\quad E_{0}=Q+m_{e};\quad E_{e}=\sqrt{m_{e}^{2}+p_{e}^{2}};$$

$$E=T_{e}=E_{e}-m_{e};\quad v_{e}=\sum_{j=1}^{3}U_{ej}v_{j}.$$

v_{e} flavor eigenstate; v_{j} mass eigenstate. The current upper limit on neutrino mass from tritium β-decay experiments holds in degenerate neutrino mass region $\left(m_{v1}\simeq m_{v2}\simeq m_{v3}\simeq\text{ with }m_{ve}=\sum_{j=1}^{3}m_{vj}/3\right)$. In the Kurie plot the Tritium decay spectrum is for a massless neutrino a straight line as a function of the electron energy. The line hits the abscissa with the electron energy at the Q-value. The Kurie plot is obtained by dividing by product $\left(K\,F\left(E,Z\right)p_{e}\,E_{e}\right)$ and taking the square root. The neutrino mass modifies the Kurie plot in the interval $<Q-m_{ve}\,c^{2}|Q=18.562\text{ keV}>$. The capture of the relic neutrinos from the CNB should show as a peak in the electron spectrum at $Q\ \ m_{ve}\ c$ The capture probability requires the same Fermi and Gamow-Teller matrix elements as for the decay. Thus the theoretical prediction for the capture probability should be as accurate as the one for the decay.

$$\Gamma^{\beta}_{capture}\left(^{3}H\right)=\frac{1}{\pi}\left(G_{F}\cos\vartheta c\right)^{2}F_{0}\left(Z+1,T_{e}\right)\left[B_{F}\left(^{3}H\right)+B_{GT}\left(^{3}H\right)\right]\rho_{e}T_{e}\times$$

$$\left\langle n_{v,e}\right\rangle\frac{n_{v,e}}{\left\langle n_{v,e}\right\rangle}$$

$$=4.2\,10^{-25}\frac{n_{v,e}}{\left\langle n_{v,e}\right\rangle}\left[\text{for one Tritium atom/ year}\right];$$

$$\text{with}:\left\langle n_{v,e}\right\rangle=56\ \text{ cm}^{-3}.$$

The values for this effective strength of the Tritium source and8 have been reduced step by step. The correct value given by Drexlin is $20\mu g$. This means 2×10^{18} Tritium$_2$ molecules. The capture rate of relic neutrinos is then:

$$\text{Capture rate at KATRIN}:\ N_{v}\left(\text{KATRIN}\right)=1.7\,10^{-6}\frac{n_{v,e}}{\left\langle n_{v,e}\right\rangle}.$$

If one uses the average relic neutrino number density $\left\langle n_{v,e}\right\rangle=56\text{cm}^{-3}$, one predict only every 590,000 years a count. But there is the hope, that the local relic neutrino density in a galaxy increases by gravitational clustering. Ringwald and Wong12 calculated, that relic neutrinos can cluster on the scale of a single galaxy and their halo and if one uses the proportionality to the baryon overdensity, then one can expect very optimistically an overdensities up to a factor $n_{v,e}/\left\langle n_{v,e}\right\rangle\leq10^{6}$

in our neighbourhood. With this very optimistic overdensity for the relic neutrinos of 106 one obtains from above equation:

$$N_\nu\left(\text{KATRIN}\right) = 1.7\,10^{-6}\frac{n_{\nu,e}}{\langle n_{\nu,e}\rangle}\left[year^{-1}\right] \approx 1.7\left[\text{counts per year}\right].$$

This seems not possible to measure for the moment. One way out would be to increase the effective activity of the tritium source. An effective mass of 2 milligrams Tritium would mean with the above optimistic estimate of the relic neutrino number overdensity $n_{\nu,e}/\langle n_{\nu,e}\rangle \approx 10^6$ about 170 counts per year, which should be feasible. But it should be possible with KATRIN, in its present form, to determine an upper limit for the local relic neutrino density of the CNB.

References

- Guth, A. H. (1998). The Inflationary Universe: The Quest for a New Theory of Cosmic Origins. Basic Books. p. 186. ISBN 978-0201328400. OCLC 35701222

- Kashlinsky (2005). "Cosmic infrared background and early galaxy evolution". Physics Reports. 409 (6): 361–438. arXiv:astro-ph/0412235. Bibcode:2005PhR...409..361K. doi:10.1016/j.physrep.2004.12.005

- Assis, A. K. T.; Neves, M. C. D. (1995). "History of the 2.7 K Temperature Prior to Penzias and Wilson" (PDF) (3): 79–87. but see also Wright, E. L. (2006). "Eddington's Temperature of Space". UCLA. Retrieved 2008-12-11

- Kogut, A.; Lineweaver, C.; Smoot, G. F.; et al. (1993). "Dipole Anisotropy in the COBE Differential Microwave Radiometers First-Year Sky Maps". Astrophysical Journal. 419: 1–6. arXiv:astro-ph/9312056. Bibcode:1993ApJ...419....1K. doi:10.1086/173453

- Hobson, M.P.; Efstathiou, G.; Lasenby, A.N. (2006). General Relativity: An Introduction for Physicists. Cambridge University Press. p. 388. ISBN 0-521-82951-8

- Gamow, G. (1948). "The Origin of Elements and the Separation of Galaxies". Physical Review. 74 (4): 505–506. Bibcode:1948PhRv...74..505G. doi:10.1103/PhysRev.74.505.2

- Nemiroff, R.; Bonnell, J., eds. (6 September 2009). "CMBR Dipole: Speeding Through the Universe". Astronomy Picture of the Day. NASA. Retrieved 18 May 2018

- Dicke, R. H.; et al. (1965). "Cosmic Black-Body Radiation". Astrophysical Journal. 142: 414–419. Bibcode:1965ApJ...142..414D. doi:10.1086/148306

Chapter 5

Dark Matter

Dark matter is a hypothetical form of matter that constitutes approximately 80% of the entire matter in the universe. The topics elaborated in this chapter include various significant topics such as cold dark matter, hot dark matter, mixed dark matter and dark matter halo, which will provide an in-depth understanding of dark matter in the universe.

In some clusters, the space between galaxies is filled with gas so hot, scientists cannot see it using visible light telescopes. The gas only can be seen as X-rays or gamma rays. Scientists look at that gas and measure how much there is between galaxies in clusters. By doing this, they discovered that there must be five times more material in the clusters than we can detect. The invisible matter that we can't detect is called "dark matter."

The Fermi Gamma-Ray Space Telescope can detect high-energy gamma rays that may be emitted when dark matter particles collide.

Perhaps the most significant sign of the existence of dark matter, however, is our very existence. Despite its invisibility, dark matter has been critical to the evolution of our universe and to the emergence of stars, planets and even life. That is because dark matter carries five times the mass of ordinary matter and, furthermore, does not directly interact with light. Both these properties were critical to the creation of structures such as galaxies—within the (relatively short) time span we know to be a typical galaxy lifetime—and, in particular, to the formation of a galaxy the size of the Milky Way. Without dark matter, radiation would have prevented clumping of the galactic structure for too long, in essence wiping it out and keeping the universe smooth and homogeneous. The galaxy essential to our solar system and our life was formed in the time since the big bang only because of the existence of dark matter.

Dark matter possibly could be brown dwarfs, "failed" stars that never ignited because they lacked the mass needed to start burning. Dark matter could be white dwarfs, the remnants of cores of dead small- to medium-size stars. Or dark matter could be neutron stars or black holes, the remnants of large stars after they explode.

However, problems exist with each of these suggestions. Scientists have strong evidence there aren't enough brown dwarfs or white dwarfs to account for all the dark matter. Black holes and neutron stars, too, are rare.

Dark matter may not be made up of the matter we are familiar with at all. The matter that makes up dark matter could different. It may be filled with particles predicted by theory but that scientists have yet to observe.

Because scientists can't see dark matter directly, they have found other ways to investigate it. We can use indirect ways to study things, like looking at a shadow and making an educated guess about what's casting the shadow. One way scientists indirectly study dark matter is by using gravitational lensing.

Light going through a gravitational lens is similar to light going through an optical lens: It gets bent. When light from distant stars passes through a galaxy or cluster, the gravity of the matter present in the galaxy or cluster causes the light to bend. As a result, the light looks like it is coming from somewhere else rather than from its actual origin. The amount of bending helps scientists learn about the dark matter present. Many NASA scientists use the Hubble Space Telescope to observe gravitational lensing.

In addition to these indirect ways, scientists at NASA think they have a direct way to detect dark matter using the Fermi Gamma-Ray Space Telescope. This telescope looks at gamma rays, the highest energy form of light. When two dark matter particles crash into each other, they might release a gamma ray. The Fermi Telescope could theoretically detect these collisions, which would appear as a burst of a gamma ray in the sky.

Technical Definition

In standard cosmology, matter is anything whose energy density scales with the inverse cube of the scale factor, i.e., $\rho \propto a^{-3}$. This is in contrast to radiation, which scales as the inverse fourth power of the scale factor $\rho \propto a^{-4}$, and a cosmological constant, which is independent of a. These scalings can be understood intuitively: for an ordinary particle in a cubical box, doubling the length of the sides of the box decreases the density (and hence energy density) by a factor of eight (2^3). For radiation, the decrease in energy density is larger because an increase in scale factor causes a proportional redshift. A cosmological constant, as an intrinsic property of space, has a constant energy density regardless of the volume under consideration.

In principle, "dark matter" means all components of the universe that are not visible but still obey $\rho \propto a^{-3}$. In practice, the term "dark matter" is often used to mean only the non-baryonic component of dark matter, i.e., excluding "missing baryons." Context will usually indicate which meaning is intended.

Observational Evidence

Galaxy Rotation Curves

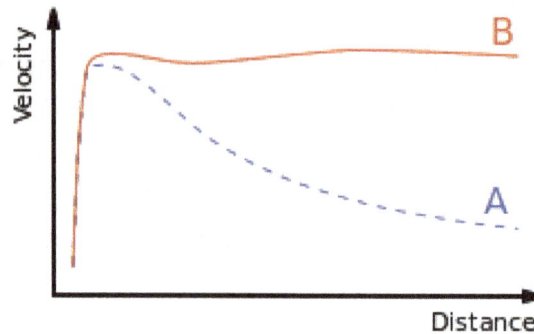

Rotation curve of a typical spiral galaxy: predicted (**A**) and observed (**B**). Dark
matter can explain the 'flat' appearance of the velocity curve out to a large radius.

The arms of spiral galaxies rotate around the galactic center. The luminous mass density of a spiral galaxy decreases as one goes from the center to the outskirts. If luminous mass were all the matter, then we can model the galaxy as a point mass in the centre and test masses orbiting around it, similar to the solar system. From Kepler's Second Law, it is expected that the rotation velocities will decrease with distance from the center, similar to the Solar System. This is not observed.Instead, the galaxy rotation curve remains flat as distance from the center increases.

If Kepler's laws are correct, then the obvious way to resolve this discrepancy is to conclude that the mass distribution in spiral galaxies is not similar to that of the Solar System. In particular, there is a lot of non-luminous matter (dark matter) in the outskirts of the galaxy.

Velocity Dispersions

Stars in bound systems must obey the virial theorem. The theorem, together with the measured velocity distribution, can be used to measure the mass distribution in a bound system, such as elliptical galaxies or globular clusters. With some exceptions, velocity dispersion estimates of elliptical galaxies do not match the predicted velocity dispersion from the observed mass distribution, even assuming complicated distributions of stellar orbits.

As with galaxy rotation curves, the obvious way to resolve the discrepancy is to postulate the existence of non-luminous matter.

Galaxy Clusters

Galaxy clusters are particularly important for dark matter studies since their masses can be estimated in three independent ways:

- From the scatter in radial velocities of the galaxies within clusters.

- From X-rays emitted by hot gas in the clusters. From the X-ray energy spectrum and flux, the gas temperature and density can be estimated, hence giving the pressure; assuming pressure and gravity balance determines the cluster›s mass profile.

- Gravitational lensing (usually of more distant galaxies) can measure cluster masses without relying on observations of dynamics (e.g., velocity).

Generally, these three methods are in reasonable agreement that dark matter outweighs visible matter by approximately 5 to 1.

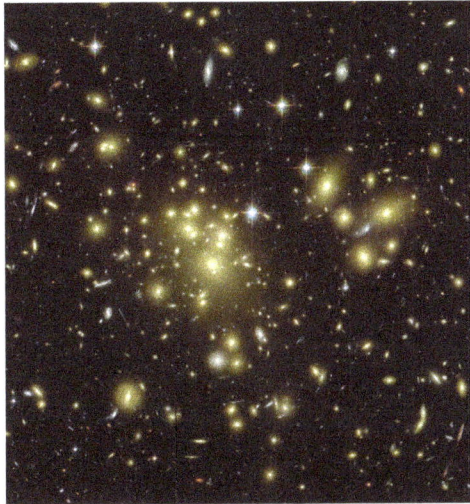

Strong gravitational lensing as observed by the Hubble Space Telescope in Abell indicates the presence of dark matter.

Gravitational Lensing

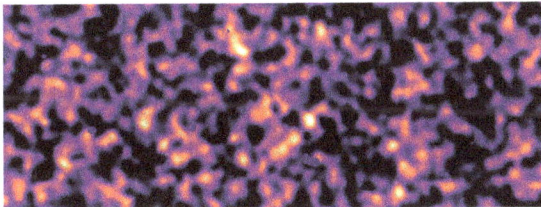

Dark matter map for a patch of sky based on gravitational lensing analysis of a Kilo-Degree survey.

One of the consequences of general relativity is that massive objects (such as a cluster of galaxies) lying between a more distant source (such as a quasar) and an observer should act as a lens to bend the light from this source. The more massive an object, the more lensing is observed.

Strong lensing is the observed distortion of background galaxies into arcs when their light passes through such a gravitational lens. It has been observed around many distant clusters including Abell 1689. By measuring the distortion geometry, the mass of the intervening cluster can be obtained. In the dozens of cases where this has been done, the mass-to-light ratios obtained correspond to the dynamical dark matter measurements of clusters. Lensing can lead to multiple copies of an image. By analyzing the distribution of multiple image copies, scientists have been able to deduce and map the distribution of dark matter around the MACS J0416.1-2403 galaxy cluster.

Weak gravitational lensing investigates minute distortions of galaxies, using statistical analyses from vast galaxy surveys. By examining the apparent shear deformation of the adjacent background galaxies, the mean distribution of dark matter can be characterized. The mass-to-light ratios correspond to dark matter densities predicted by other large-scale structure measurements. Dark matter does not bend light itself; mass (in this case the mass of the dark matter) bends spacetime. Light follows the curvature of spacetime, resulting in the lensing effect.

Cosmic Microwave Background

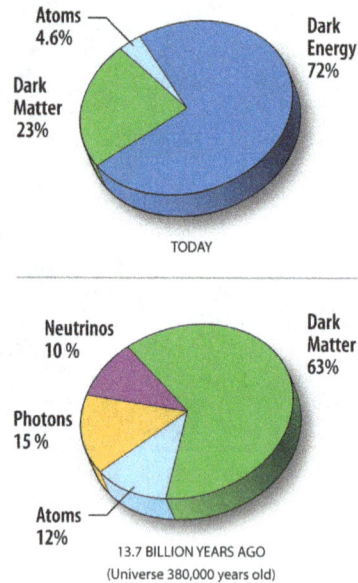

Estimated division of total energy in the universe into matter, dark matter and dark energy based on five years of WMAP data.

Although both dark matter and ordinary matter are matter, they do not behave in the same way. In particular, in the early universe, ordinary matter was ionized and interacted strongly with radiation via Thomson scattering. Dark matter does not interact directly with radiation, but it does affect the CMB by its gravitational potential (mainly on large scales), and by its effects on the density and velocity of ordinary matter. Ordinary and dark matter perturbations, therefore, evolve differently with time and leave different imprints on the cosmic microwave background (CMB).

The cosmic microwave background is very close to a perfect blackbody but contains very small temperature anisotropies of a few parts in 100,000. A sky map of anisotropies can be decomposed into an angular power spectrum, which is observed to contain a series of acoustic peaks at near-equal spacing but different heights. The series of peaks can be predicted for any assumed set of cosmological parameters by modern computer codes such as CMBFast and CAMB, and matching theory to data, therefore, constrains cosmological parameters. The first peak mostly shows the density of baryonic matter, while the third peak relates mostly to the density of dark matter, measuring the density of matter and the density of atoms.

The CMB anisotropy was first discovered by COBE in 1992, though this had too coarse resolution to detect the acoustic peaks. After the discovery of the first acoustic peak by the balloon-borne BOOMERanGexperiment in 2000, the power spectrum was precisely observed by WMAP in 2003-12, and even more precisely by the Planck spacecraft in 2013-15. The results support the Lambda-CDM model.

The observed CMB angular power spectrum provides powerful evidence in support of dark matter, as its precise structure is well fitted by the Lambda-CDM model but difficult to reproduce with any competing model such as MOND.

Structure Formation

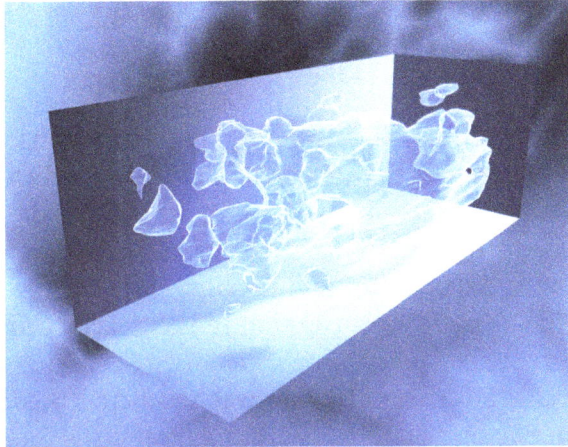

3D map of the large-scale distribution of dark matter, reconstructed from measurements of weak gravitational lensing with the Hubble Space Telescope.

Structure formation refers to the period after the Big Bang when density perturbations collapsed to form stars, galaxies, and clusters. Prior to structure formation, the Friedmann solutions to general relativity describe a homogeneous universe. Later, small anisotropies gradually grew and condensed the homogeneous universe into stars, galaxies and larger structures. Ordinary matter is affected by radiation, which is the dominant element of the universe at very early times. As a result, its density perturbations are washed out and unable to condense into structure. If there were only ordinary matter in the universe, there would not have been enough time for density perturbations to grow into the galaxies and clusters currently seen.

Dark matter provides a solution to this problem because it is unaffected by radiation. Therefore, its density perturbations can grow first. The resulting gravitational potential acts as an attractive potential well for ordinary matter collapsing later, speeding up the structure formation process.

Bullet Cluster

If dark matter does not exist, then the next most likely explanation is that general relativity – the prevailing theory of gravity – is incorrect. The Bullet Cluster, the result of a recent collision of two galaxy clusters, provides a challenge for modified gravity theories because its apparent center of mass is far displaced from the baryonic center of mass. Standard dark matter theory can easily explain this observation, but modified gravity has a much harder time, especially since the observational evidence is model-independent.

Type Ia Supernova Distance Measurements

Type Ia supernovae can be used as standard candles to measure extragalactic distances, which can in turn be used to measure how fast the universe has expanded in the past. The data indicates that the universe is expanding at an accelerating rate, the cause of which is usually ascribed to dark energy. Since observations indicate the universe is almost flat, it is expected that the total energy density of everything in the universe to sum to 1 ($\Omega_{tot} \sim 1$). The measured dark energy density is $\Omega_\Lambda = \sim 0.690$; the observed ordinary (baryonic) matter energy density is $\Omega_b = \sim 0.0482$ and the energy density of radiation is negligible. This leaves a missing $\Omega_{dm} = \sim 0.258$ that nonetheless behaves like matter – dark matter.

Sky Surveys and Baryon Acoustic Oscillations

Baryon acoustic oscillations (BAO) are regular, periodic fluctuations in the density of the visible baryonic matter (normal matter) of the universe. These are predicted to arise in the Lambda-CDM model due to the early universe›s acoustic oscillations in the photon-baryon fluid and can be observed in the cosmic microwave background angular power spectrum. BAOs set up a preferred length scale for baryons. As the dark matter and baryons clumped together after recombination, the effect is much weaker in the galaxy distribution in the nearby universe, but is detectable as a subtle ($\sim$ 1 percent) preference for pairs of galaxies to be separated by 147 Mpc, compared to those separated by 130 or 160 Mpc. This feature was predicted theoretically in the 1990s and then discovered in 2005, in two large galaxy redshift surveys, the Sloan Digital Sky Survey and the 2dF Galaxy Redshift Survey. Combining the CMB observations with BAO measurements from galaxy redshift surveys provides a precise estimate of the Hubble constant and the average matter density in the Universe. The results support the Lambda-CDM model.

Redshift-space Distortions

Large galaxy redshift surveys may be used to make a three-dimensional map of the galaxy distribution. These maps are slightly distorted because distances are estimated from observed redshifts; the redshift contains a contribution from the galaxy's so-called peculiar velocity in addition to the dominant Hubble expansion term. On average, superclusters are expanding but more slowly than the cosmic mean due to their gravity, while voids are expanding faster than average. In a redshift map, galaxies in front of a supercluster have excess radial velocities towards it and have redshifts slightly higher than their distance would imply, while galaxies behind the supercluster have redshifts slightly low for their distance. This effect causes superclusters to appear squashed in the radial direction, and likewise voids are stretched. Their angular positions are unaffected. The effect is not detectable for any one structure since the true shape is not known, but can be measured by averaging over many structures assuming Earth is not at a special location in the Universe.

The effect was predicted quantitatively by Nick Kaiser in 1987, and first decisively measured in 2001 by the 2dF Galaxy Redshift Survey. Results are in agreement with the Lambda-CDM model.

Lyman-alpha Forest

In astronomical spectroscopy, the Lyman-alpha forest is the sum of the absorption lines arising from the Lyman-alphatransition of neutral hydrogen in the spectra of distant galaxies and quasars. Lyman-alpha forest observations can also constrain cosmological models. These constraints agree with those obtained from WMAP data.

Composition of Dark Matter: Baryonic vs. Nonbaryonic

Dark matter can refer to any substance that interacts predominantly via gravity with visible matter (e.g., stars and planets). Hence in principle it need not be composed of a new type of fundamental particle but could, at least in part, be made up of standard baryonic matter, such as protons or neutrons. However, for the reasons outlined below, most scientists think the dark matter is dominated by a non-baryonic component, which is likely composed of a currently unknown fundamental particle (or similar exotic state).

Baryonic Matter

Baryons (protons and neutrons) make up ordinary stars and planets. However, baryonic matter also encompasses less common black holes, neutron stars, faint old white dwarfs and brown dwarfs, collectively known as massive compact halo objects(MACHOs), which can be hard to detect.

However multiple lines of evidence suggest the majority of dark matter is not made of baryons:

- Sufficient diffuse, baryonic gas or dust would be visible when backlit by stars.

- The theory of Big Bang nucleosynthesis predicts the observed abundance of the chemical elements. If there are more baryons, then there should also be more helium, lithium and heavier elements synthesized during the Big Bang. Agreement with observed abundances requires that baryonic matter makes up between 4–5% of the universe›s critical density. In contrast, large-scale structure and other observations indicate that the total matter density is about 30% of the critical density.

- Astronomical searches for gravitational microlensing in the Milky Way found that at most a small fraction of the dark matter may be in dark, compact, conventional objects (MACHOs, etc.); the excluded range of object masses is from half the Earth›s mass up to 30 solar masses, which covers nearly all the plausible candidates.

- Detailed analysis of the small irregularities (anisotropies) in the cosmic microwave background. Observations by WMAP and Planck indicate that around five-sixths of the total matter is in a form that interacts significantly with ordinary matter or photons only through gravitational effects.

Non-baryonic Matter

Candidates for non-baryonic dark matter are hypothetical particles such as axions, sterile neutrinos, weakly-interacting massive particles (WIMPs), gravitationally-interacting massive particles (GIMPs), or supersymmetric particles. The three neutrino types already observed are indeed abundant, and dark, and matter, but because their individual masses – however uncertain they may be – are almost certainly tiny, they can only supply a small fraction of dark matter, due to limits derived from large-scale structure and high-redshift galaxies.

Unlike baryonic matter, nonbaryonic matter did not contribute to the formation of the elements in the early universe (Big Bang nucleosynthesis) and so its presence is revealed only via its gravitational effects, or, weak lensing. In addition, if the particles of which is composed are supersymmetric, they can undergo annihilation interactions with themselves, possibly resulting in observable by-products such as gamma rays and neutrinos (indirect detection).

Dark Matter Aggregation and Dense Dark Matter Objects

If dark matter is as common as observations suggest, an obvious question is whether it can form objects equivalent to planets, stars, or black holes. The answer has historically been that it cannot, because of two factors:

- *It lacks an efficient means to lose energy:* Ordinary matter forms dense objects because it has numerous ways to lose energy. Losing energy would be essential for object formation, because a particle that gains energy during compaction or falling «inward» under gravity, and cannot lose it any other way, will heat up and increase velocity and momentum. Dark matter appears to lack means to lose energy, simply because it is not capable of interacting strongly in other ways except through gravity. The Virial theorem suggests that such a particle would not stay bound to the gradually forming object - as the object began to form and compact, the dark matter particles within it would speed up and tend to escape.

- *It lacks a range of interactions needed to form structures:* Ordinary matter interacts in many different ways. This allow it to form more complex structures. For example, stars form through gravity, but the particles within them interact and can emit energy in the form of neutrinos and electromagnetic radiation through fusion when they become energetic enough. Protons and neutrons can bind via the strong interaction and then form atoms with electrons largely through electromagnetic interaction. But there is no evidence that dark matter is capable of such a wide variety of interactions, since it only seems to interact through gravity and through some means no stronger than the weak interaction. (although this is speculative until dark matter is better understood).

This question has been debated heavily during recent years. In 2016–2017 the idea of dense dark matter or dark matter being black holes, including primordial black holes, made a comeback following results of gravitation wave detection. These were again ruled out in December 2017, but research and theories based on these still continue as at 2018, including approaches to dark matter cooling, and the question is by no means settled.

Classification of Dark Matter: Cold, Warm or Hot

Dark matter can be divided into *cold*, *warm*, and *hot* categories. These categories refer to velocity rather than an actual temperature, indicating how far corresponding objects moved due to random motions in the early universe, before they slowed due to cosmic expansion – this is an important distance called the free streaming length (FSL). Primordial density fluctuations smaller than this length get washed out as particles spread from overdense to underdense regions, while larger fluctuations are unaffected; therefore this length sets a minimum scale for later structure formation. The categories are set with respect to the size of a protogalaxy (an object that later evolves into a dwarf galaxy): dark matter particles are classified as cold, warm, or hot according as their FSL; much smaller (cold), similar (warm), or much larger (hot) than a protogalaxy.

Mixtures of the above are also possible: a theory of mixed dark matter was popular in the mid-1990s, but was rejected following the discovery of dark energy.

Cold dark matter leads to a bottom-up formation of structure while hot dark matter would result in a top-down formation scenario; the latter is excluded by high-redshift galaxy observations.

Alternative Definitions

These categories also correspond to fluctuation spectrum effects and the interval following the Big Bang at which each type became non-relativistic. Davis *et al.* wrote in 1985:

> Candidate particles can be grouped into three categories on the basis of their effect on the fluctuation spectrum(Bond *et al.* 1983). If the dark matter is composed of abundant light particles which remain relativistic until shortly before recombination, then it may be termed "hot". The best candidate for hot dark matter is a neutrino. A second possibility is for the dark matter particles to interact more weakly than neutrinos, to be less abundant, and to have a mass of order 1 keV. Such particles are termed "warm dark matter", because they have lower thermal velocities than massive neutrinos there are at present few candidate particles which fit this description. Gravitinos and photinos have been suggested (Pagels and Primack 1982; Bond, Szalay and Turner 1982). Any particles which became nonrelativistic very early, and so were able to diffuse a negligible distance, are termed "cold" dark matter (CDM). There are many candidates for CDM including supersymmetric particles.

> — *M. Davis, G. Efstathiou, C. S. Frenk, and S. D. M. White*

Another approximate dividing line is that warm dark matter became non-relativistic when the universe was approximately 1 year old and 1 millionth of its present size and in the radiation-dominated era (photons and neutrinos), with a photon temperature 2.7 million K. Standard physical cosmology gives the particle horizon size as $2ct$ (speed of light multiplied by time) in the radiation-dominated era, thus 2 light-years. A region of this size would expand to 2 million light years today (absent structure formation). The actual FSL is approximately 5 times the above length, since it continues to grow slowly as particle velocities decrease inversely with the scale factor after they become non-relativistic. In this example the FSL would correspond to 10 million light-years or 3 Mpc today, around the size containing an average large galaxy.

The 2.7 million K photon temperature gives a typical photon energy of 250 electron-volts, thereby setting a typical mass scale for warm dark matter: particles much more massive than this, such as GeV – TeV mass WIMPs, would become non-relativistic much earlier than 1 year after the Big Bang and thus have FSLs much smaller than a protogalaxy, making them cold. Conversely, much lighter particles, such as neutrinos with masses of only a few eV, have FSLs much larger than a protogalaxy, thus qualifying them as hot.

Cold Dark Matter

Cold dark matter offers the simplest explanation for most cosmological observations. It is dark matter composed of constituents with an FSL much smaller than a protogalaxy. This is the focus for dark matter research, as hot dark matter does not seem capable of supporting galaxy or galaxy cluster formation, and most particle candidates slowed early.

The constituents of cold dark matter are unknown. Possibilities range from large objects like MA-CHOs (such as black holes) or RAMBOs (such as clusters of brown dwarfs), to new particles such as WIMPs and axions.

Studies of Big Bang nucleosynthesis and gravitational lensing convinced most cosmologists that MACHOs cannot make up more than a small fraction of dark matter. According to A. Peter: "... the only *really plausible* dark-matter candidates are new particles."

The 1997 DAMA/NaI experiment and its successor DAMA/LIBRA in 2013, claimed to directly detect dark matter particles passing through the Earth, but many researchers remain skeptical, as negative results from similar experiments seem incompatible with the DAMA results.

Many supersymmetric models offer dark matter candidates in the form of the WIMPy Lightest Supersymmetric Particle(LSP). Separately, heavy sterile neutrinos exist in non-supersymmetric extensions to the standard model that explain the small neutrino mass through the seesaw mechanism.

Warm Dark Matter

Warm dark matter comprises particles with an FSL comparable to the size of a protogalaxy. Predictions based on warm dark matter are similar to those for cold dark matter on large scales, but with less small-scale density perturbations. This reduces the predicted abundance of dwarf galaxies and may lead to lower density of dark matter in the central parts of large galaxies. Some researchers consider this a better fit to observations. A challenge for this model is the lack of particle candidates with the required mass ~ 300 eV to 3000 eV.

No known particles can be categorized as warm dark matter. A postulated candidate is the sterile neutrino: a heavier, slower form of neutrino that does not interact through the weak force, unlike other neutrinos. Some modified gravity theories, such as scalar-tensor-vector gravity, require "warm" dark matter to make their equations work.

Hot Dark Matter

Hot dark matter consists of particles whose FSL is much larger than the size of a protogalaxy. The neutrino qualifies as such particle. They were discovered independently, long before the hunt for dark matter: they were postulated in 1930, and detected in 1956. Neutrinos' mass is less than 10^{-6} that of an electron. Neutrinos interact with normal matter only via gravity and the weak force, making them difficult to detect (the weak force only works over a small distance, thus a neutrino triggers a weak force event only if it hits a nucleus head-on). This makes them 'weakly-interacting light particles' (WILPs), as opposed to WIMPs.

The three known flavours of neutrinos are the *electron*, *muon*, and *tau*. Their masses are slightly different. Neutrinos oscillate among the flavours as they move. It is hard to determine an exact upper bound on the collective average mass of the three neutrinos (or for any of the three individually). For example, if the average neutrino mass were over 50 eV/c^2 (less than $10^{-5}$of the mass of an electron), the universe would collapse. CMB data and other methods indicate that their average mass probably does not exceed 0.3 eV/c^2. Thus, observed neutrinos cannot explain dark matter.

Because galaxy-size density fluctuations get washed out by free-streaming, hot dark matter implies that the first objects that can form are huge supercluster-size pancakes, which then fragment into galaxies. Deep-field observations show instead that galaxies formed first, followed by clusters and superclusters as galaxies clump together.

Detection of Dark Matter Particles

If dark matter is made up of sub-atomic particles, then millions, possibly billions, of such particles must pass through every square centimeter of the Earth each second. Many experiments aim to test this hypothesis. Although WIMPs are popular search candidates, the Axion Dark Matter eXperiment (ADMX) searches for axions. Another candidate is heavy hidden sector particles that only interact with ordinary matter via gravity.

These experiments can be divided into two classes: direct detection experiments, which search for the scattering of dark matter particles off atomic nuclei within a detector; and indirect detection, which look for the products of dark matter particle annihilations or decays.

Direct Detection

Direct detection experiments aim to observe low-energy recoils (typically a few keVs) of nuclei induced by interactions with particles of dark matter, which (in theory) are passing through the Earth. After such a recoil the nucleus will emit energy as, e.g., scintillation light or phonons, which is then detected by sensitive apparatus. To do this effectively it is crucial to maintain a low background, and so such experiments operate deep underground to reduce the interference from cosmic rays. Examples of underground laboratories with direct detection experiments include the Stawell mine, the Soudan mine, the SNOLAB underground laboratory at Sudbury, the Gran Sasso National Laboratory, the Canfranc Underground Laboratory, the Boulby Underground Laboratory, the Deep Underground Science and Engineering Laboratory and the China Jinping Underground Laboratory.

These experiments mostly use either cryogenic or noble liquid detector technologies. Cryogenic detectors operating at temperatures below 100mK, detect the heat produced when a particle hits an atom in a crystal absorber such as germanium. Noble liquid detectors detect scintillation produced by a particle collision in liquid xenon or argon. Cryogenic detector experiments include: CDMS, CRESST, EDELWEISS, EURECA. Noble liquid experiments include ZEPLIN, XENON, DEAP, ArDM, WARP, DarkSide, PandaX, and LUX, the Large Underground Xenon experiment. Both of these techniques focus strongly on their ability to distinguish background particles (which predominantly scatter off electrons) from dark matter particles (that scatter off nuclei). Other experiments include SIMPLE and PICASSO.

Currently there has been no well-established claim of dark matter detection from a direct detection experiment, leading instead to strong upper limits on the mass and interaction cross section with nucleons of such dark matter particles. The DAMA/NaI and more recent DAMA/LIBRA experimental collaborations claim to have detected an annual modulation in the rate of events in their detectors, which they claim is due to dark matter. This results from the expectation that as the Earth orbits the Sun, the velocity of the detector relative to the dark matter halo will vary by a small amount. This claim is so far unconfirmed and in contradiction with negative results from other experiments such as LUX and SuperCDMS.

A special case of direct detection experiments covers those with directional sensitivity. This is a search strategy based on the motion of the Solar System around the Galactic Center. A low pressure time projection chamber makes it possible to access information on recoiling tracks and constrain WIMP-nucleus kinematics. WIMPs coming from the direction in which the Sun travels (approximately towards Cygnus) may then be separated from background, which should be isotropic. Directional dark matter experiments include DMTPC, DRIFT, New-age and MIMAC.

Indirect Detection

Collage of six cluster collisions with dark matter maps. The clusters were
observed in a study of how dark matter in clusters of galaxies
behaves when the clusters collide.

Indirect detection experiments search for the products of the self-annihilation or decay of dark matter particles in outer space. For example, in regions of high dark matter density (e.g., the centre of our galaxy) two dark matter particles could annihilate to produce gamma rays or Standard Model particle-antiparticle pairs.Alternatively if the dark matter particle is unstable, it could decay into standard model (or other) particles. These processes could be detected indirectly through an excess of gamma rays, antiprotons or positrons emanating from high density regions in our galaxy or others. A major difficulty inherent in such searches is that various astrophysical sources can mimic the signal expected from dark matter, and so multiple signals are likely required for a conclusive discovery.

A few of the dark matter particles passing through the Sun or Earth may scatter off atoms and lose energy. Thus dark matter may accumulate at the center of these bodies, increasing the chance of collision/annihilation. This could produce a distinctive signal in the form of high-energy neutrinos. Such a signal would be strong indirect proof of WIMP dark matter. High-energy neutrino telescopes such as AMANDA, IceCube and ANTARES are searching for this signal. The detection by LIGO in September 2015 of gravitational waves, opens the possibility of observing dark matter in a new way, particularly if it is in the form of primordial black holes.

Many experimental searches have been undertaken to look for such emission from dark matter annihilation or decay, examples of which follow. The Energetic Gamma Ray Experiment Telescope observed more gamma rays in 2008 than expected from the Milky Way, but scientists concluded that this was most likely due to incorrect estimation of the telescope›s sensitivity.

The Fermi Gamma-ray Space Telescope is searching for similar gamma rays. In April 2012, an analysis of previously available data from its Large Area Telescope instrument produced statistical evidence of a 130 GeV signal in the gamma radiation coming from the center of the Milky Way. WIMP annihilation was seen as the most probable explanation.

At higher energies, ground-based gamma-ray telescopes have set limits on the annihilation of dark matter in dwarf spheroidal galaxies and in clusters of galaxies.

The PAMELA experiment (launched 2006) detected excess positrons. They could be from dark matter annihilation or from pulsars. No excess antiprotons were observed.

In 2013 results from the Alpha Magnetic Spectrometer on the International Space Station indicated excess high-energy cosmic rays that could be due to dark matter annihilation.

Collider Searches for Dark Matter

An alternative approach to the detection of dark matter particles in nature is to produce them in a laboratory. Experiments with the Large Hadron Collider (LHC) may be able to detect dark matter particles produced in collisions of the LHC protonbeams. Because a dark matter particle should have negligible interactions with normal visible matter, it may be detected indirectly as (large amounts of) missing energy and momentum that escape the detectors, provided other (non-negligible) collision products are detected. Constraints on dark matter also exist from the LEP experiment using a similar principle, but probing the interaction of dark matter particles with electrons rather than quarks. It is important to note that any discovery from collider searches must be corroborated by discoveries in the indirect or direct detection sectors to prove that the particle discovered is, in fact, the dark matter of our Universe.

Alternative Hypotheses

Because dark matter remains to be conclusively identified, many other hypotheses have emerged aiming to explain the observational phenomena that dark matter was conceived to explain. The most common method is to modify general relativity. General relativity is well-tested on solar system scales, but its validity on galactic or cosmological scales has not been well proven. A suitable modification to general relativity can conceivably eliminate the need for dark matter. The most well-known theories of this class are MOND and its relativistic generalization TeVeS, f(R) gravity and entropic gravity. Alternative theories abound.

A problem with alternative hypotheses is that the observational evidence for dark matter comes from so many independent approaches. Explaining any individual observation is possible but explaining all of them is very difficult. Nonetheless, there have been some scattered successes for alternative hypotheses, such as a 2016 test of gravitational lensing in entropic gravity.

The prevailing opinion among most astrophysicists is that while modifications to general relativity can conceivably explain part of the observational evidence, there is probably enough data to conclude there must be some form of dark matter.

Baryonic Dark Matter

The properties of a fermionic dark matter candidate which has baryon number and refer to this type of scenario as "Baryonic Dark Matter." We show the relic density constraints and the predictions for direct detection experiments. Since this model has only four free parameters one could hope to test this idea combining the possible results from the Large Hadron Collider and dark matter experiments.

Recently, it proposed a simple extension of the Standard Model where one can understand the spontaneous breaking of baryon and lepton numbers at the low scale. Here, will be discussed a simplified version of this model, only considering baryon number as a local gauge symmetry. Therefore, this model is based on the gauge group

$$SU(3)_C \otimes SU(2)_L \otimes U(1)_Y \otimes U(1)_B.$$

In order to define an anomaly-free theory using this gauge group, we need to include additional fermions that account for anomaly cancellation,

$$\Psi_L \sim (1,2,-\frac{1}{2},B_1), \qquad \Psi_R \sim (1,2,-\frac{1}{2},B_2),$$

$$\eta_R \sim (1,1,-1,B_1), \qquad \eta_L \sim (1,1,-1,B_2),$$

$$\chi_R \sim (1,1,0,B_1), \qquad \chi_L \sim (1,1,0,B_2),$$

and extend the scalar sector with a new Higgs boson to allow for a spontaneous breaking of baryon number,

$$S_B \sim (1,1,0,-3).$$

Here B_1 and B_2 refer to the baryon numbers of the additional fermions which are vector-like under the SM gauge group. From the conditions that ensure the cancellation of all relevant baryonic anomalies, one finds the following relation for the baryon numbers of the new fermions:

$$B_1 - B_2 = -3$$

The relevant interactions of the new fields in the theory are

$$-\mathcal{L} \supset \lambda_1 \bar{\Psi}_L \Psi_R S_B + \lambda_2 \bar{\eta}_R \eta_L S_B + \lambda_3 \bar{\chi}_R \chi_L S_B + h.c.$$

Notice that one can have terms such as $a_1 \chi_L \chi_L S_B$ and $a_2 \chi_R \chi_R S_B^{\dagger}$ B only when $B_1 = -B_2$. Here we will stick to the case where $B_1 \neq -B_2$. The Yukawa interactions between the new fields and the Standard Model Higgs boson are present as well, but they are not relevant for our main discussion. It is important to notice that the new Higgs boson must have baryon number -3 in order to generate vector-like mass for the new fermions. Therefore, once S_B gets a vacuum expectation value breaking local $U(1)_B$ we never generate any operator mediating proton decay and the scale for baryon number violation can be as low as a few TeV.

In this simple theory, when the local baryon number is spontaneously broken by the vacuum expectation value v_B of S_B, one obtains a Z_2 discrete symmetry which protects the stability of the dark matter candidate. Under this Z_2 the new fermionic fields transform as

$$\Psi_{L,R} \rightarrow -\Psi_{L,R}, \; \eta_{L,R,} \rightarrow -\eta_{L,R}, \text{and } \chi_{L,R} \rightarrow -\chi_{L,R}.$$

Therefore, when the lightest new field with baryon number is neutral, one can have a candidate for the cold dark matter in the Universe. For simplicity, we will focus on the case when the dark matter is SM singlet-like and is the Dirac fermion $\chi = \chi_R + \chi_L$. Since the dark matter has baryon number, the relevant interactions with the new gauge boson Z_B related to baryon number are

$$\mathcal{L} \supset g_B \, \overline{\chi} \gamma_\mu Z_B^\mu \, (B_2 \, P_L + B_1 \, P_R \,) \chi \,,$$

where P_L and P_R are the left- and right-handed projectors, and g_B is the gauge coupling related to baryon number. Of course, the new gauge boson also couples to the SM quarks, which is crucial to understand the properties of the dark matter candidate. The leptophobic gauge boson mass reads as

$$M_{Z_B} = 3 g_B v_B,$$

while the mass of the SM singlet-like baryonic dark matter candidate is given by

$$M_\chi = \lambda_3 v_B / \sqrt{2} < \frac{\sqrt{2\pi}}{3} \frac{M_{Z_B}}{g_B}$$

This upper limit is coming from the perturbative condition on the Yukawa coupling λ_3, i.e. $|\lambda_3|^2/4\pi < 1$. It is important to notice that this model for baryonic dark matter has only four free parameters:

$$M_\chi, M_{Z_B}, g_B, and \; B,$$

and one needs to satisfy the relic density constraints and the bounds from direct detection. Here $B = B_1 + B_2$ is the parameter which enters in the predictions for the relevant cross sections. One could imagine that the parameters M_{Z_B} and g_B can be determined from the discovery of a leptophobic gauge boson at the Large Hadron Collider. Therefore, one can say that for a given value of these two quantities we can find the values of B and M_χ using the relic density and spin-independent cross section values. We will discuss in more details the numerical predictions in the following sections in order to appreciate this connection between collider physics and dark matter experiments. In the rest of the paper we will neglect the kinetic mixing between $U(1)_B$ and $U(1)_Y$, as well as the mixing between the SM Higgs and S_B.

Non-Baryonic Dark Matter

Non-baryonic matter is matter that, unlike all the kind kinds of matter with which we are familiar, is not made of baryons (including the neutrons and protons found in all atomic nuclei). Proposed as a possible constituent of dark matter, it could come in two forms, classified as cold non-baryonic matter or hot non-baryonic matter. The former would be made of particles moving much slower than light, of which there are several, as yet undetected, candidates; hot non-baryonic matter would be made of particles moving very fast, such as neutrinos. Non-baryonic matter (hot or cold) is supposed to interact weakly with radiation. Therefore, the imprints left by the non-baryonic matter in the cosmic background radiation would be different than those left by the baryonic matter. This attribute could be used to measure the contribution of non-baryonic matter to the total amount of mass in the universe.

Cold Dark Matter

CDM has long been the leading candidate for what this missing mass in the universe is. However, some researchers still favor a combination theory, where aspects of all three types of dark matter exist together to make up the total missing mass.

CDM is a kind of dark matter that, if it exists, moves slowly compared to the speed of light. It is thought to have been present in the universe since the very beginning and has very likely influenced the growth and evolution of galaxies. as well as the formation of the first stars. Astronomers and physicists think that it's most likely some exotic particle that hasn't yet been detected. It very likely has some very specific properties:

It would have to lack an interaction with the electromagnetic force. This is fairly obvious, since dark matter is dark. Therefore it doesn't interact with, reflect, or radiate any type of energy in the electromagnetic spectrum.

However, any candidate particle that makes up cold dark matter would have to take into account that it has to interact with a gravitational field. For proof of this, astronomers have noticed that dark matter accumulations in galaxy clusters wield a gravitational influence on light from more distant objects that happens to be passing by. This so-called "gravitational lensing effect" has been observed many times.

Candidate Cold Dark Matter Objects

While no known matter meets all of the criteria for cold dark matter, at least three theories have been advanced to explain CDM.

- *Weakly Interacting Massive Particles*: Also known as WIMPs, these particles, by definition, meet all the needs of CDM. However, no such particle has ever been found to exist. WIMPs have become the catch all term for all cold dark matter candidates, regardless of why the particle is thought to arise.

- *Axions*: These particles possess (at least marginally) the necessary properties of dark matter, but for various reasons are probably not the answer to the question of cold dark matter.

- *MACHOs*: This is an acronym for *Massive Compact Halo Objects*, which are objects like black holes, ancient neutron stars, brown dwarfs and planetary objects. These are all non-luminous and massive. But, because of their large sizes, both in terms of volume and mass, they would be relatively easy to detect by monitoring localized gravitational interactions. However, there are problems with the MACHO hypothesis. The observed motion of galaxies, for instance, is uniform in a way that would be hard to explain if MACHOs supplied the missing mass. Furthermore, star clusters would require a very uniform distribution of such objects within their boundaries. That seems very unlikely. Also, the sheer number of MACHOs that would have to be fairly large in order to explain the missing mass.

Structure Formation

In the cold dark matter theory, structure grows hierarchically, with small objects collapsing under their self-gravity first and merging in a continuous hierarchy to form larger and more massive objects. In the hot dark matter paradigm, popular in the early 1980s, structure does not form hierarchically (*bottom-up*), but rather forms by fragmentation (*top-down*), with the largest superclusters forming first in flat pancake-like sheets and subsequently fragmenting into smaller pieces like our galaxy the Milky Way. Predictions of the cold dark matter paradigm are in general agreement with observations of cosmological large scale structure.

Lambda CDM Model

Since the late 1980s or 1990s, most cosmologists favor the cold dark matter theory (specifically the modern Lambda-CDM model) as a description of how the Universe went from a smooth initial state at early times to the lumpy distribution of galaxies and their clusters we see today — the large-scale structure of the Universe. The theory sees the role that dwarf galaxies played as crucial, as they are thought to be natural building blocks that form larger structures, created by small-scale density fluctuations in the early Universe.

Hot Dark Matter

Hot DM refers to particles, such as neutrinos, that were moving at nearly the speed of light at redshift $z \sim 10^6$ (or time $t \sim 1\,\text{yr}$), when the temperature $T \sim 3 \times 10^2$ eV and the cosmic horizon first encompassed 10^{12} $M\odot$, the amount of dark matter contained in the halo of a large galaxy like the Milky Way. Hot DM particles must also be still in thermal equilibrium after the last phase transition in the hot early universe, the QCD confinement transition, which presumably took place at $T_{QCD} \approx 10^2$ MeV. Hot DM particles have a cosmological number density roughly comparable to that of the microwave background photons, which implies an upper bound to their mass of a few tens of eV. This then implies that free streaming of these relativistic particles destroys any fluctuations smaller than supercluster size, $\sim 10^{15}$ $M\odot$.

The main problem with HDM theory is that the high speeds of the particles (i.e. neutrinos) in the early universe could not have allowed small density fluctuations to clump together in order to create the large fluctuations we see now. We believe matter (or in other words galaxies) is distributed throughout the universe as it is now due to the growth of small initial fluctuations. Since neutrinos would have been moving so fast that these tiny initial fluctuations would have been smoothed out, HDM theory cannot account for the distribution of galaxies in the universe. The small scale of this initial clumping that is impossible for neutrinos to maintain is supported by COBE observations.

Galaxy Formation with Hot DM

The standard hot DM candidate is massive neutrinos, although other, more exotic theoretical possibilities have been suggested, such as a "majoron" of nonzero mass which is lighter than the lightest neutrino species, and into which all neutrinos decay. Neutrinos appeared to be an attractive DM candidate because of the measurement of an electron neutrino mass of about 30 eV in 1980 . This coincided with the improving CMB limits on the primordial fluctuation amplitude, which forced Zel'dovich and other theorists to abandon the idea that all the dark matter could be made of ordinary baryonic matter. The version of HDM that they worked out in detail, with adiabatic Gaussian primordial fluctuations, became the prototype for the subsequent $\Omega_m = 1$ CDM theory.

Mass Constraints

Direct measurements of neutrino masses have given only upper limits. A secure upper limit on the electron neutrino mass is roughly 15 eV. The Particle Data Group notes that a more precise limit cannot be given since unexplained effects have resulted in significantly negative measurements of $m(v_e)^2$ in tritium beta decay experiments. However, this problem is at least partially resolved, and the latest experimental upper limits on the electron neutrino mass are 2.8 eV from the Mainz and 2.5 eV from the Troitsk tritium beta decay experiments (both 95% C.L.). There is an upper limit on an effective Majorana neutrino mass of ~ 1eV from neutrinoless double beta decay experiments . The upper limits from accelerator experiments on the masses of the other neutrinos are $m(v_\mu) < 0.17 \, \mathrm{MeV} \, (90\% \, \mathrm{CL})$ and $m(v_\tau) < 18 \, \mathrm{MeV} \, (95\% \, \mathrm{CL})$, but since stable neutrinos with such large masses would certainly "overclose the universe" (i.e., contribute such a large cosmological density that the universe could never have attained its present age), cosmology implies a much lower upper limit on these neutrino masses.

Before going further, it will be necessary to discuss the thermal history of neutrinos in the standard hot big bang cosmology in order to derive the corresponding constraints on their mass. Left-handed neutrinos of mass ≤ 1 MeV remain in thermal equilibrium until the temperature drops to T_{vd}, at which point their mean free path first exceeds the horizon size and they essentially cease interacting thereafter, except gravitationally. Their mean free path is, in natural units

$(\hbar = c = 1)$, $\lambda_v \sim [\sigma_v n_{e\pm}]^{-1} [(G_F^2 T^2)(T^3)]^{-1}$, where $G_F \approx 10^{-5} \, G_e \, \mathrm{V}^{-2}$ is the Fermi constant that measures the strength of the weak interactions. The horizon size is $\lambda_v \sim (G\rho)^{-1/2} \sim M_{P\ell} T^{-2}$,

the Planck mass $M_{P\ell} \equiv G^{-1/2} = 1.22 \times 10^{19} \mathrm{GeV}$ Thus $\lambda_h / \lambda_v \sim (T/T_{vd})^3$, with the neutrino decoupling temperature

$$T_{vd} \sim M_{P\ell}^{-1/3} G_F^{-1/3} \sim 1 \mathrm{MeV}.$$

After T drops below $\frac{1}{2}$ MeV, e^+e^- annihilation ceases to be balanced by pair creation, and the entropy of the e^+e^- pairs heats the photons. Above 1 MeV, the number density $n_{v,i}$ of each left-handed neutrino species and its right-handed antiparticle is equal to that of the photons, n_γ, times the factor 3/4 from Fermi versus Bose statistics. But then e^+e^- annihilation increases the photon number density relative to that of the neutrinos by a factor of 11/4. As a result, the neutrino temperature $T_{v,0} = (4/11)^{1/3} T_{\gamma,0}$. Thus today, for each species,

$$n_{v,0} = \frac{3}{4} \cdot \frac{4}{11} n_{\gamma,0} = 109\,\theta^3\ \mathrm{cm}^{-3},$$

where $\theta \equiv (T_0/2.7\,\mathrm{K})$. With the cosmic background radiation temperature $T_0 = 2.728 \pm 0.004$ K measured by the FIRAS instrument on the COBE satellite, $T_{v,0} = 1.947$ K and $n_{v,0} = 112\,\mathrm{cm}^{-3}$.

Since the present cosmological matter density is

$$\bar{\rho}m = \Omega \rho_c = 10.54\,\Omega_m h^2\ \mathrm{ke\,V\,cm}^{-3},$$

it follows that

$$\sum_i m_{vi} < \bar{\rho}m/n_{v,0} \leq 96\,\Omega_m h^2 \theta^{-3}\ \mathrm{eV} \approx 93\Omega_m h^2\ \mathrm{eV},$$

where the sum runs over all neutrino species with $M_{vi} \leq 1$ MeV. Observational data imply that $\Omega_m h^2 \approx 0.1 - 0.3$ since $\Omega_m \approx 0.3 - 0.5$ and $h \approx 0.65 \pm 0.1$. Thus if all the dark matter were light neutrinos, the sum of their masses would be $\approx 9 - 28$ eV.

In deriving above equation, we have been assuming that all the neutrino species are light enough to still be relativistic at decoupling, i.e. lighter than an MeV. The bound shows that they must then be much lighter than that. In the alternative case that a neutrino species is nonrelativistic at decoupling, it has been shown that its mass must then exceed several GeV, which is not true of the known neutrinos $(v_e, v_\mu$ and $v_\tau)$. (One might at first think that the Boltzmann factor would sufficiently suppress the number density of neutrinos weighing a few tens of MeV to allow compatibility with the present density of the universe. It is the fact that they "freeze out" of equilibrium well before the temperature drops to their mass that leads to the higher mass limit.) We have also been assuming that the neutrino chemical potential is negligible, i.e. that $|n_v - n_{\bar{v}}| \ll n_\gamma$. This is very plausible, since the net baryon number density $(n_b - n_{\bar{b}}) \leq 10^{-9}\ n_\gamma$, and big bang nucleosynthesis restricts the allowed parameters

Phase Space Constraint

We have just seen that light neutrinos must satisfy an upper bound on the sum of their masses. But now we will discuss a lower bound on neutrino mass that arises because they must be rather massive to form the dark matter in galaxies, since their phase space density is limited by the Pauli exclusion principle. A slightly stronger bound follows from the fact that they were not degenerate in the early universe.

The phase space constraint follows from Jeans's theorem in classical mechanics to the effect that the maximum 6-dimensional phase space density cannot increase as a system of collisionless particles evolves. At early times, before density inhomogenitites become nonlinear, the neutrino phase space density is given by the Fermi-Dirac distribution

$$n_v(p) = \frac{g_v}{h^3}\left[1 + \exp\left(\frac{pc}{kT_v(z)}\right)\right]^{-1},$$

where here h is Planck's constant and $g_\nu = 2$ for each species of left-handed ν plus right-handed $\bar{\nu}$. Since momentum and temperature both scale as redshift z as the universe expands, this distribution remains valid after neutrinos drop out of thermal equilibrium at ~ 1 MeV, and even into the nonrelativistic regime $T_\nu < m_\nu$. The standard version of the phase space constraint follows from demanding that the central phase space density $9[2(2\pi)^{5/2} G r_c^2 \sigma m_\nu^4]^{-1}$ of the DM halo, assumed to be an isothermal sphere of core radius rc and one-dimensional velocity dispersion σ, not exceed the maximum value of the initial phase space density $n_\nu(0) = g_\nu/2h^3$. The result is

$$m_\nu > (120\,eV)\left(\frac{100\,\mathrm{km\,s^{-1}}}{\sigma}\right)^{1/4}\left(\frac{1\,\mathrm{kpc}}{r_c}\right)^{1/2}\left(\frac{g_\nu}{2}\right)^{-1/4}.$$

The strongest lower limits on mv follow from applying this to the smallest galaxies. Both theoretical arguments regarding the dwarf spheroidal (dS) satellite galaxies of the Milky Way and data on Draco, Carina, and Ursa Minor made it clear some time ago that dark matter dominates the gravitational potential of these dS galaxies, and the case has only strengthened with time. The phase space constraint then sets a lower limit $m_\nu > 500$ eV, which is completely incompatible with the cosmological constraint. However, this argument only excludes neutrinos as the DM in certain small galaxies; it remains possible that the DM in these galaxies is (say) baryonic, while that in larger galaxies such as our own is (at least partly) light neutrinos. A more conservative phase space constraint was obtained for the Draco and Ursa Minor dwarf spheroidals, but the authors concluded that neutrinos consistent with the cosmological upper bound on mv cannot be the DM in these galaxies. A similar analysis applied to the gas-rich lowrotation-velocity dwarf irregular galaxy DDO 154 gave a limit mv > 94 eV, again inconsistent with the cosmological upper bound.

Free Streaming

The most salient feature of hot DM is the erasure of small fluctuations by free streaming. Thus even collisionless particles effectively exhibit a Jeans mass. It is easy to see that the minimum mass of a surviving fluctuation is of order M_{Pl}^3/m_ν^2. Let us suppose that some process in the very early universe — for example, thermal fluctuations subsequently vastly inflated in the inflationary scenario — gave rise to adiabatic fluctuations on all scales. In adiabatic fluctuations, all the components — radiation and matter — fluctuate together. Neutrinos of nonzero mass mv stream relativistically from decoupling until the temperature drops to $T \sim m_\nu$, during which time they traverse a distance $d_\nu = R_H\,(T = m_\nu) \sim M_{Pl}\,m_\nu^{-2}$. In order to survive this free streaming, a neutrino fluctuation must be larger in linear dimension than dv. Correspondingly, the minimum mass in neutrinos of a surviving fluctuation is $M_{J,\nu} \sim d_\nu^3 m_\nu n_\nu(T = m_\nu) \sim d_\nu^3 m_\nu^4 \sim M_{Pl}^3 m_\nu^{-2}$. By analogy with Jeans's calculation of the minimum mass of an ordinary fluid perturbation for which gravity can overcome pressure, this is referred to as the (free-streaming) Jeans mass.

A more careful calculation gives

$$d_\nu = 41(m_\nu/30\,\mathrm{eV})^{-1}(1+z)^{-1}\,\mathrm{Mpc},$$

that is, $d_\nu = 41(m_\nu/30\,\mathrm{eV})^{-1}$ Mpc in comoving coordinates, and correspondingly

$$M_{J,\nu} = 1.77 M_{Pl}^3 \, m_\nu^{-2} = 3.2 \times 10^{15} \, (m_\nu / 30\,\text{eV})^{-2} \, M_\odot,$$

which is the mass scale of superclusters. Objects of this size are the first to form in a ν-dominated universe, and smaller scale structures such as galaxies can form only after the initial collapse of supercluster-size fluctuations. When a fluctuation of total mass $\sim 10^{15} \, M_\odot$ enters the horizon at $z \sim 10^4$, the density contrast δ_{RB} of the radiation plus baryons ceases growing and instead starts oscillating as an acoustic wave, while that of the massive neutrinos δ_ν continues to grow linearly with the scale factor $R = (1 + z)^{-1}$ since the Compton drag that prevents growth of δRB does not affect the neutrinos.

By recombination, at $Z_r \sim 10^3$, $\delta_{RB}/\delta_\nu \lesssim \sim 10^{-1}$, with possible additional suppression of δRB by Silk damping. Thus the hot DM scheme with adiabatic primordial fluctuations predicts small-angle fluctuations in the microwave background radiation that are lower than in the adiabatic baryonic cosmology, which was one of the reasons HDM appealed to Zel'dovich and other theorists. Similar considerations apply in the warm and cold DM schemes. However, as we will discuss in a moment, the HDM top-down sequence of cosmogony is wrong, and with the COBE normalization hardly any structure would form by the present.

In numerical simulations of dissipationless gravitational clustering starting with a fluctuation spectrum appropriately peaked at $\lambda \sim d\nu$ (reflecting damping by free streaming below that size and less time for growth of the fluctuation amplitude above it), the regions of high density form a network of filaments, with the highest densities occurring at the intersections and with voids in between. The similarity of these features to those seen in observations was cited as evidence in favor of HDM.

Problems with ν DM

A number of potential problems with the neutrino dominated universe had emerged by about 1983, however.

- From studies both of nonlinear clustering (comoving length scale $\lambda \leq 10$ Mpc) and of streaming velocities in the linear regime ($\lambda > 10$ Mpc), it follows that supercluster collapse must have occurred recently: $z_{sc} \leq 0.5$ is indicated and in any case $z_{sc} < 2$. However, the best limits on galaxy ages coming from globular clusters and other stellar populations indicated that galaxy formation took place before $z \approx 3$. Moreover, if quasars are associated with galaxies, as is suggested by the detection of galactic luminosity around nearby quasars and the apparent association of more distant quasars with galaxy clusters, the abundance of quasars at $z > 2$ was also inconsistent with the "top-down" neutrino dominated scheme in which superclusters form first: $z_{sc} > z_{galaxies}$.

- Numerical simulations of the nonlinear "pancake" collapse taking into account dissipation of the baryonic matter showed that at least 85% of the baryons are so heated by the associated shock that they remain unable to condense, attract neutrino halos, and eventually form galaxies. This was a problem for the hot DM scheme for two reasons. With the primordial nucleosynthesis constraint $\Omega b \ll 0.1$, there would be difficulty having enough baryonic matter condense to form the luminosity that we actually observe. And, where are the X-rays from the shock-heated pancakes.

- The neutrino picture predicts that there should be a factor of ~ 5 increase in M/M_b between large galaxies ($M \sim 10^{12}\, M_\odot$) and large clusters ($M \geq 10^{14}\, M_\odot$), since the larger clusters, with their higher escape velocities, are able to trap a considerably larger fraction of the neutrinos. Although there is some indication that the mass-to-light ratio M/L increases with M, the ratio of total to luminous mass M/Mlum is probably a better indicator of the value of M/Mb, and it is roughly the same for galaxies with large halos and for rich clusters.

These problems, while serious, would perhaps not have been fatal for the hot DM scheme. But an even more serious problem for HDM arose from the low amplitude of the CMB fluctuations detected by the COBE satellite, $(\Delta T/T)_{rms} = (1.1 \pm 0.2) \times 10^{-5}$ smoothed on an angular scale of about 10°. Although HDM and CDM both have the Zel'dovich spectrum shape ($P(k) \propto k$) in the long-wavelength limit, because of the free-streaming cutoff the amplitude of the HDM spectrum must be considerably higher in order to form any structure by the present. With the COBE normalization, the HDM spectrum is only beginning to reach nonlinearity at the present epoch. Thus the evidence against standard hot DM is convincing. At very least, it indicates that structure formation in a neutrino-dominated universe must be rather more complicated than in the standard inflationary picture. The main alternative that has been considered is cosmic strings plus hot dark matter. Because the strings would continue to seed structure up until the present, and because these seeds are in the nature of rather localized fluctuations, hot DM would probably work better with string seeds than cold DM. However, strings and other cosmic defect models are now essentially ruled out because they predict that the cosmic microwave background would have an angular power spectrum without the pronounced (doppler/acoustic/Sakharov) peak at angular wavenumber $l \sim 220$ that now appears to be clearly indicated by the data, along with secondary peaks at higher l.

Mixed Dark Matter

Mixed dark matter, i.e. a mixture of cold and hot dark matter, became fashionable for a few years in the mid-1990. The main virtue of this scenario was that it could reconcile the CMBR observations of the time with (apparently spurious) reports of neutrino masses in the 20–30 eV range. Interest in mixed dark matter declined, however, once the LCDM model entered the stage and provided a superior explanation for supernova type Ia data, the CMBR anisotropies and the observed large scale structure of the Universe at both high and low redshift. While neutrinos are no longer believed to contribute substantially to the energy density of the Universe, they do appear to have non-zero masses in the right range to make them act as HDM. A small contribution from HDM (with $\Omega_{HDM} \sim 0.01$) is still viable within the LCDM picture, and could help to lower the central densitites of dark halos somewhat.

The Lyman$-\alpha$ forest constraint on the WDM mass in MDM models is milder than that in pure WDM models. The structure formation in MDM models is not well established even after some previous efforts. MDM models also attract interests in the context of the reported anomalous X-ray line in stacked X-ray spectra in XMM-Newton and Chandra data. While the anomaly has not been confirmed in Suzaku data, the 3.5 keV unidentified X-ray line may originate from the decay of sterile neutrinos. Harada and Kamada show that decaying 7 keV sterile neutrinos that are produced via non-resonant process called Dodelson-Widrow mechanism can reproduce the 3.5 keV X-ray line if they account for 20–60% of the present mass density of DM. Interestingly this MDM model can also mitigate the missing satellite problem while evading constraints from the Lyman $-\alpha$ forests.

Dark Matter Halo

A dark halo is the inferred halo of invisible material (dark matter) that permeates and surrounds individual galaxies, as well as groups and clusters of galaxies.

Evidence for the existence of dark matter first came from studies of the motions of stars and gas in galaxies. For example, the quantity and distribution of luminous matter within disk galaxies cannot account for the rotation curves observed, implying a significant invisible component. In a similar manner, velocity dispersions measured in the outer regions of elliptical galaxies are higher than expected given the luminous matter within the galaxy. Estimates based on these considerations suggest that perhaps as much as 90% of the matter present in galaxies is in the form of dark matter.

Evidence for dark matter is also found through observations of the motions of galaxies in groups and clusters. Using a similar argument to that applied to stellar motions within galaxies, the velocities of galaxies in groups and clusters are so high that the group or cluster would fly apart if only the luminous matter were present.

In addition, the X-ray emitting gas often found throughout galaxy clusters indicates that these clusters must contain large amounts of dark matter. With a temperature in excess of a million degrees, this gas would evaporate from the cluster if the visible elements were the only components. In both galaxies and groups and clusters of galaxies, the dark matter is found to be distributed in a roughly spherical halo around the visible component – the dark halo. In the Milky Way the dark halo appears to extend out to at least 300,000 light years, and possibly even further, reaching far beyond the extent of the visible matter in the disk.

Rotation Curves as Evidence of a Dark Matter Halo

The presence of dark matter (DM) in the halo is inferred from its gravitational effect on a spiral galaxy's rotation curve. Without large amounts of mass throughout the (roughly spherical) halo, the rotational velocity of the galaxy would decrease at large distances from the galactic center, just as the orbital speeds of the outer planets decrease with distance from the Sun. However, observations of spiral galaxies, particularly radio observations of line emission from neutral atomic hydrogen (known, in astronomical parlance, as HI), show that the rotation curve of most spiral galaxies flattens out, meaning that rotational velocities do not decrease with distance from the galactic center. The absence of any visible matter to account for these observations implies either that unobserved ("dark") matter, first proposed by Ken Freeman in 1970, exist, or that the theory of motion under gravity (General Relativity) is incomplete. Freeman noticed that the expected decline in velocity was not present in NGC 300 nor M33, and considered an undetected mass to explain it. The DM Hypothesis has been reinforced by several studies.

Formation and Structure of Dark Matter Halos

The formation of dark matter halos is believed to have played a major role in the early formation of galaxies. During initial galactic formation, the temperature of the baryonic matter should have still been much too high for it to form gravitationally self-bound objects, thus requiring the prior formation of dark matter structure to add additional gravitational interactions. The current hypothesis for this is based on cold dark matter (CDM) and its formation into structure early in the universe.

The hypothesis for CDM structure formation begins with density perturbations in the Universe that grow linearly until they reach a critical density, after which they would stop expanding and collapse to form gravitationally bound dark matter halos. These halos would continue to grow in mass (and size), either through accretion of material from their immediate neighborhood, or by merging with other halos. Numerical simulations of CDM structure formation have been found to proceed as follows: A small volume with small perturbations initially expands with the expansion of the Universe. As time proceeds, small-scale perturbations grow and collapse to form small halos. At a later stage, these small halos merge to form a single virialized dark matter halo with an ellipsoidal shape, which reveals some substructure in the form of dark matter sub-halos.

The use of CDM overcomes issues associated with the normal baryonic matter because it removes most of the thermal and radiative pressures that were preventing the collapse of the baryonic matter. The fact that the dark matter is cold compared to the baryonic matter allows the DM to form these initial, gravitationally bound clumps. Once these subhalos formed, their gravitational interaction with baryonic matter is enough to overcome the thermal energy, and allow it to collapse into the first stars and galaxies. Simulations of this early galaxy formation matches the structure observed by galactic surveys as well as observation of the Cosmic Microwave Background.

Density Profiles

A commonly used model for galactic dark matter halos is the pseudo-isothermal halo:

$$\rho(r) = \rho_o \left[1 + \left(\frac{r}{r_c} \right)^2 \right]^{-1}$$

where ρ_o denotes the finite central density and r_c the core radius. This provides a good fit to most rotation curve data. However, it cannot be a complete description, as the enclosed mass fails to converge to a finite value as the radius tends to infinity. The isothermal model is, at best, an approximation. Many effects may cause deviations from the profile predicted by this simple model. For example, (i) collapse may never reach an equilibrium state in the outer region of a dark matter halo, (ii) non-radial motion may be important, and (iii) mergers associated with the (hierarchical) formation of a halo may render the spherical-collapse model invalid.

Numerical simulations of structure formation in an expanding universe lead to the theoretical prediction of the NFW (Navarro-Frenk-White) profile:

$$\rho(r) = \frac{\rho_{crit}\delta_c}{\left(\dfrac{r}{r_s}\right)\left(1+\dfrac{r}{r_s}\right)^2}$$

where r_s is a scale radius, δ_c is a characteristic (dimensionless) density, and $\rho_{crit} = 3H^2/8\pi G$ is the critical density for closure. The NFW profile is called 'universal' because it works for a large variety of halo masses, spanning four orders of magnitude, from individual galaxies to the halos of galaxy clusters. This profile has a finite gravitational potential even though the integrated mass still diverges logarithmically. It has become conventional to refer to the mass of a halo at a fiducial point that encloses an overdensity 200 times greater than the critical density of the universe, though mathematically the profile extends beyond this notational point. It was later deduced that the density profile depends on the environment, with the NFW appropriate only for isolated halos. NFW halos generally provide a worse description of galaxy data than does the pseudo-isothermal profile, leading to the cuspy halo problem.

Higher resolution computer simulations are better described by the Einasto profile:

$$\rho(r) = \rho_e Exp\left[-d_n\left(\left(\frac{r}{r_c}\right)^{\frac{1}{n}}-1\right)\right]$$

where r is the spatial (i.e., not projected) radius. The term d_n is a function of n such that ρ_e is the density at the radius r_e that defines a volume containing half of the total mass. While the addition of a third parameter provides a slightly improved description of the results from numerical simulations, it is not observationally distinguishable from the 2 parameter NFW halo, and does nothing to alleviate the cuspy halo problem.

Shape

The collapse of overdensities in the cosmic density field is generally aspherical. So, there is no reason to expect the resulting halos to be spherical. Even the earliest simulations of structure formation in a CDM universe emphasized that the halos are substantially flattened. Subsequent work has shown that halo equidensity surfaces can be described by ellipsoids characterized by the lengths of their axes.

Because of uncertainties in both the data and the model predictions, it is still unclear whether the halo shapes inferred from observations are consistent with the predictions of ΛCDM cosmology.

Halo Substructure

Up until the end of the 1990s, numerical simulations of halo formation revealed little substructure. With increasing computing power and better algorithms, it became possible to use greater numbers of particles and obtain better resolution. Substantial amounts of substructure are now expected. When a small halo merges with a significantly larger halo it becomes a subhalo orbiting within the potential well of its host. As it orbits, it is subjected to strong tidal forces from the host, which cause it to lose mass. In addition the orbit itself evolves as the subhalo is subjected to dynamical friction which causes it to lose energy and angular momentum to the dark matter particles of its host. Whether a subhalo survives as a self-bound entity depends on its mass, density profile, and its orbit.

Angular Momentum

As originally pointed out by Hoyle and first demonstrated using numerical simulations by Efstathiou & Jones, asymmetric collapse in an expanding universe produces objects with significant angular momentum.

Numerical simulations have shown that the spin parameter distribution for halos formed by dissipation-less hierarchical clustering is well fit by a log-normal distribution, the median and width of which depend only weakly on halo mass, redshift, and cosmology:

$$\rho(\lambda)d\lambda = \frac{1}{\sqrt{2\pi}\sigma_{ln\lambda}}\exp\left[-\frac{\ln\left(\frac{\lambda}{\bar{\lambda}}\right)^2}{2\sigma_{\ln\lambda}^2}\right]\frac{d\lambda}{\lambda}$$

with $\bar{\lambda} \approx 0.035$ and $\sigma_{ln\lambda} \approx 0.5$. At all halo masses, there is a marked tendency for halos with higher spin to be in denser regions and thus to be more strongly clustered.

Theories about the Nature of Dark Matter

The nature of dark matter in the galactic halos of spiral galaxies is still undetermined, but there are two popular theories: either the halo is composed of weakly interacting elementary particles known as WIMPs, or it is composed of a number of small, dark bodies known as MACHOs. MACHOs, an acronym for massive compact halo objects, would be composed of "ordinary" matter that simply does not emit easily detectable radiation. The reliance of modern astronomy on detecting electromagnetic radiation would render dim, but massive objects, nearly undetectable. A wide range of possible MACHO candidates have been suggested, encompassing everything from black holes to very dim dwarf stars. Some have hypothesised that the dark matter halo consists of primordial black holes. Despite being too dim to detect directly via electromagnetic telescopes, MACHOs would necessarily interact gravitationally, as described by general relativity. The preferred method for searching for MACHOs in the halo of our own galaxy has been to look for microlensing events. Gravitational microlensing occurs when two stars fall on a common line of sight, rendering the far star obscured by the near one. However, as light from the far star passes through the gravitational well of the near star the light bends, creating an Einstein Halo. In a microlensing event, the halo is so small that it is optically indistinguishable from the star. The overall effect is to simply make the star appear brighter. The EROS and MACHO projects search of MACHOs in the halo by observing stars in the small and large Magellanic clouds. If a MACHO existed in the halo along the line of sight of stars in the cloud, the lensing would brighten it from that orientation as opposed to others. The magnitude and number (or lack of) lensing events can be used to place bounds on the masses of any MACHOs which might be in the halo. The two projects initially were able to place a very strict limit on MACHOS in the range of $3.5 \cdot 10^{-7} M_\odot < m < 4.5 \cdot 10^{-5} M_\odot$, concluding that such light mass MACHOs could, at most, only constitute 10% of the accepted mass of our dark matter halo. Two years later, the EROS2 project extended this limit, concluding that any MACHO less than a single solar mass could not make up a significant part of the halo. The two collaborations were able to extend this to a rather large window, ruling out any MACHO in the $3.5^{-7} M_\odot < m < 30 M_\odot$ window. The final window of extremely heavy MACHOs above $43 M_\odot$ were ruled out by comparing Monte Carlo method simulations to observed distributions, which the authors indicate as "The End of the MACHO Era". Extremely light MACHOs less than the current limit are also not a viable possibility, as such light mass objects would not have survived on the timescale that it would take for the galaxy to form.

Dark Halo Around the Milky Way

The Milky Way's dark halo is believed to outweigh the galaxy's normal matter by around a factor of 20. While the inner edge of the luminous hypothesized outer ring that surrounds the spiral disk of the Milky Way may be around 120,000 light-years (ly) across, the dark halo encompasses and permeates even the enormous luminous halo of scattered individual stars and globular clusters, extending some 300,000 to 400,000 ly out from the galactic center in radius (twice that in diameter). In 2006, a team of scientists (including Jürg Diemand, Michael Kuhlen, and Piero Madau) modelled the process by which dark matter "clouds" are attracted to form the Milky Way's dark halo (Diemand et al, 2006). They simulated the development and movement of 234 million "cloudlets" as they come together to form a dark halo about the same size as that around the Milky Way. Their simulations show that there should be at least 10,000 separate "subhaloes" of dark matter within the overall galactic halo, each at least a few thousand light years across. Over time, a fair number of these galactic "seeds" should have attracted ordinary matter (mostly hydrogen and helium gas) to form star clusters. About 120 of the larger clumps of dark matter should have become large enough to have attracted enough gas to become dwarf galaxies, although astronomers have identified about 15 dwarf satellite galaxies around the Milky Way thus far.

The Milky Way's bright ordinary matter is composed of a spiral disk and outer ring,
which in turn is embedded in a larger luminous halo that is only
the visible part of an even larger halo of dark matter.

It is possible that many subhaloes did not form dwarf galaxies because dark matter has some property that prevents it from forming dense clumps. For example, it might be unexpectedly hot, and therefore hard to compress. Many astronomers currently believe, however, that there are other explanations for the paucity of observed satellite companions around the Milky Way. It may be that most of the subhaloes were sterilized by ultraviolet light from the earliest stars, which heated up intergalactic gas so that that it has been more difficult for subhaloes to capture. In addition, supernova explosions may have blasted gas out of many early dwarf galaxies, halting their continued development. In November 2009, two astronomers revealed supporting evidence that Smith's Cloud may be surrounded by a massive dark halo.

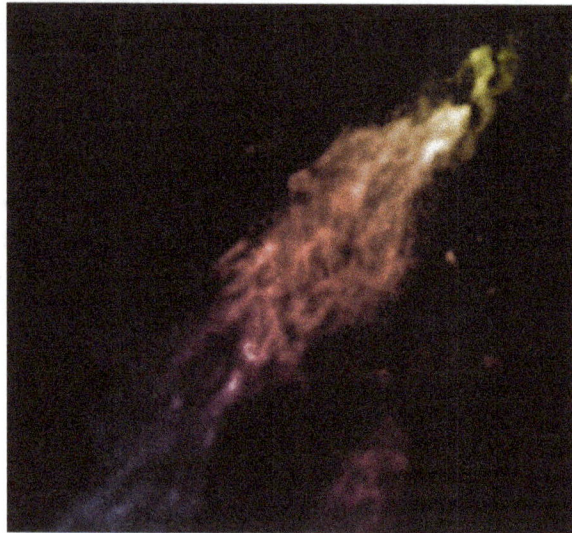

Smith's Cloud, a massive gas cloud that is apparently surrounded by an even
more massive dark halo, will collide with the disk of the Milky
Way galaxy within 20-40 million years.

Astronomers have detected evidence that dwarf satellite galaxies are disturbing the cocoon of dark matter around the Milky Way and causing its disk to warp. In 1957, astronomers surveying galactic hydrogen gas discovered that the Milky Way is not flat but warped near its edges like a fedora hat, with one side of its spiral disk curving as much as 20,000 light years above the main galactic plane and the other dipping a little less below it. While some researchers suspected the warp was caused by the two Magellanic Clouds (nearby satellite galaxies that orbit the Milky Way every 1.5 billion years), subsequent calculations showed they alone were not massive enough to produce the disk's warp. In January 2006, a team of researchers (including Leo Blitz, Martin Weinberg, Carl Heiles, and Evan Levine) announced finding evidence that the Magellanic Clouds can account for the warp but only because their motion around the Milky Way generates a powerful gravitational wake within the massive dark halo. As the Magellanic Clouds orbit the Milky Way, computer simulations indicate that the galactic disk ripples over time and its edges ruffle "like a table cloth in the breeze". On January 9, 2007, astronomers announced that new measurements of the velocities of the Megallanic Clouds through space suggest that the Milky Way's combined dark and visible mass must be twice as much as originally thought if the Clouds are trully orbiting satellites of the galaxy.

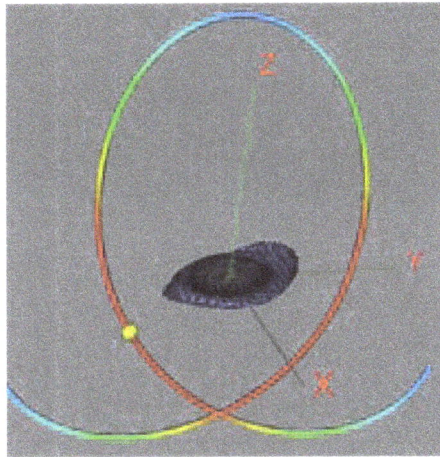

The Milky Way's disk is warped from the orbital movement and interaction
of the Magellanic Clouds through the galaxy's dark matter halo.

In January 2010, a team of astronomers (including David Law) announced at the 215th Meeting of the American Astronomical Society that the cloud of dark matter that surrounds our Milky Way galaxy appears to be shaped like a "squashed beach ball" that is oriented perpendicularly to the galaxy's spiral disk. The team studied the path of a dwarf galaxy called the Sagittarius dwarf elliptical or spheroidal galaxy, whose stars have been "shredded" into a long tidal stream as the smaller galaxy was gravitationally drawn into a merger with the Milky Way beginning some three billion years ago. Their hypothesis is that the the grivitional pull of the Milky Way's immense halo of surrounding dark matter should have shaped the trajectory of the tidal stream of stars ripped from the Sagittarius galaxy as it was drawn in.

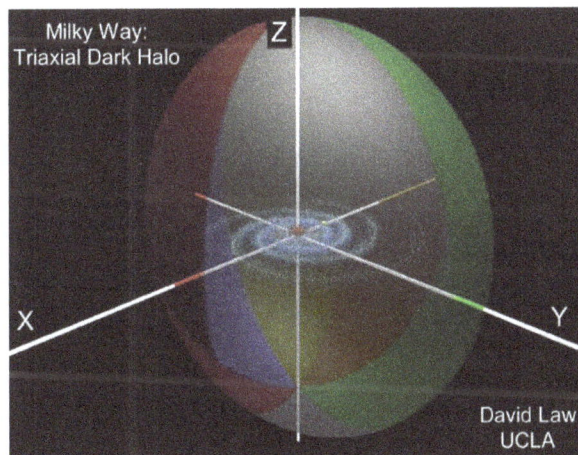

The Milky Way's halo of dark matter is oriented roughly perpendicularly to the spiral disk in
the shape of a squashed beach ball -- Sol's location is markeded by the yellow dot.

Their study of this "debris" stream of stars suggests that the distribution of dark matter around the Milky Way is very different to that of the galaxy's stars and gas matter seen in luminous ordinary matter. Although computer simulations had suggested that the halo should mimick the Milky Way's spiral disk of stars, the team's results indicate that the dark halo is oriented roughly perpendicular to the disk and is distributed roughly half as thick as it is wide.

The "tidal stream" of a small galaxy merging into the Milky Way (SagDEG), detected using red (M) giant stars from the Two-Micron All Sky Survey (2MASS), was analyzed to suggest the shape of the Milky Way's halo of dark matter.

References

- Freese, Katherine (4 May 2014). The Cosmic Cocktail: Three Parts Dark Matter. Princeton University Press. ISBN 978-1-4008-5007-5

- Roberts, Morton S.; Whitehurst, Robert N. (October 1975). "The rotation curve and geometry of M31 at large galactocentric distances". The Astrophysical Journal. 201: 327–346. Bibcode:1975ApJ...201..327R. doi:10.1086/153889

- Overbye, Dennis (December 27, 2016). "Vera Rubin, 88, Dies; Opened Doors in Astronomy, and for Women". The New York Times. Retrieved December 27, 2016

- Rogstad, D. H.; Shostak, G. Seth (September 1972). "Gross Properties of Five Scd Galaxies as Determined from 21-centimeter Observations". The Astrophysical Journal. 176: 315–321. Bibcode:1972ApJ...176..315R. doi:10.1086/151636

- Theo Koupelis; Karl F Kuhn (2007). In Quest of the Universe. Jones & Bartlett Publishers. p. 492; Figure 16-13. ISBN 0-7637-4387-9

- Roberts, Morton S. (May 1966). "A High-Resolution 21-cm Hydrogen-Line Survey of the Andromeda Nebula". The Astrophysical Journal. 159: 639–656. Bibcode:1966ApJ...144..639R. doi:10.1086/148645

- Some details of Zwicky's calculation and of more modern values are given in Richmond, M., Using the virial theorem: the mass of a cluster of galaxies, retrieved 10 July 2007

- Taylor, A. N.; et al. (1998). "Gravitational Lens Magnification and the Mass of Abell 1689". The Astrophysical Journal. 501 (2): 539–553. arXiv:astro-ph/9801158. Bibcode:1998ApJ...501..539T. doi:10.1086/305827

- Mark H. Jones; Robert J. Lambourne; David John Adams (2004). An Introduction to Galaxies and Cosmology. Cambridge University Press. p. 21; Figure 1.13. ISBN 0-521-54623-0

- Stonebraker, Alan (3 January 2014). "Synopsis: Dark-Matter Wind Sways through the Seasons". Physics – Synopses. American Physical Society. Retrieved 6 January 2014

- Bergstrom, L. (2000). "Non-baryonic dark matter: Observational evidence and detection methods". Reports on Progress in Physics. 63 (5): 793–841. arXiv:hep-ph/0002126. Bibcode:2000RPPh...63..793B. doi:10.1088/0034-4885/63/5/2r3

Permissions

Index

www.ingramcontent.com/pod-product-compliance
Lightning Source LLC
Chambersburg PA
CBHW082026190326

41458CB00010B/3292